T0155520

SpringerBriefs in Optimization

Series Editors

Sergiy Butenko
Mirjam Dür
Panos M. Pardalos
János D. Pintér
Stephen M. Robinson
Tamás Terlaky
My T. Thai

SpringerBriefs in Optimization showcases algorithmic and theoretical techniques, case studies, and applications within the broad-based field of optimization. Manuscripts related to the ever-growing applications of optimization in applied mathematics, engineering, medicine, economics, and other applied sciences are encouraged.

More information about this series at http://www.springer.com/series/8918

Balázs Bánhelyi • Tibor Csendes • Balázs Lévai
László Pál • Dániel Zombori

The GLOBAL Optimization Algorithm

Newly Updated with Java Implementation
and Parallelization

 Springer

Balázs Bánhelyi
Department of Computational Optimization
University of Szeged
Szeged, Hungary

Balázs Lévai
NNG Inc
Szeged, Hungary

Dániel Zombori
Department of Computational Optimization
University of Szeged
Szeged, Hungary

Tibor Csendes
Department of Computational Optimization
University of Szeged
Szeged, Hungary

László Pál
Sapientia Hungarian University
of Transylvania
Miercurea Ciuc
Romania

ISSN 2190-8354   ISSN 2191-575X   (electronic)
SpringerBriefs in Optimization
ISBN 978-3-030-02374-4   ISBN 978-3-030-02375-1   (eBook)
https://doi.org/10.1007/978-3-030-02375-1

Library of Congress Control Number: 2018961407

Mathematics Subject Classification: 90-08, 90C26, 90C30

This Springer imprint is published by the registered company Springer Nature Switzerland AG
The registered company address is: Gewerbestrasse 11, 6330 Cham, Switzerland

Acknowledgments

The authors are grateful to their families for the patience and support that helped to produce the present volume.

This research was supported by the project "Integrated program for training new generation of scientists in the fields of computer science," EFOP-3.6.3-VEKOP-16-2017-0002. The project has been supported by the European Union, co-funded by the European Social Fund, and by the János Bolyai Research Scholarship of the Hungarian Academy of Sciences.

Contents

Chapter 1
Introduction

1.1 Introduction

In our modern days, working on global optimization [25, 27, 34, 38, 51, 67] is not the privilege of academics anymore, these problems surround us, they are present in our daily life through the increasing number of smart devices to mention one of the most obvious evidences, but they also affect our life when public transport or shipping routes are determined, or the placement of new disposal sites are decided, to come up with some less obvious examples.

Although there are still a lot of open problems in classic mathematical fields [3], like circle packing or covering (e.g., [39, 68]), dynamical systems [5, 13, 14], the list of natural [2], life [4, 42, 41], and engineering fields of science, which yield new and new problems to solve, is practically infinite. Simultaneously, a good optimization tool becomes more and more valuable especially if it is flexible and capable of addressing a wide range of problems.

GLOBAL is a stochastic algorithm [12, 15] aiming to solve bound constrained, nonlinear, global optimization problems. It was the first available implementation of the stochastic global optimization algorithm of Boender et al. [11], which attracted several publications in those years and later [6, 7, 8, 9, 10, 37, 53, 59, 60]. Although, it was competitive and efficient compared to other algorithms, it has not been improved much in the last decade. Its latest implementations were available in FORTRAN, C, and Matlab. Knowing its potential (its competitiveness was documented recently in [15]), we decided to modernize this algorithm to provide the scientific community a whole framework that offers easy customization with more options and better performance than its predecessors.

© The Author(s), under exclusive licence to Springer International Publishing AG, part of Springer Nature 2018
B. Bánhelyi et al., *The GLOBAL Optimization Algorithm*,
SpringerBriefs in Optimization, https://doi.org/10.1007/978-3-030-02375-1_1

1.2 Problem Domain

In this book, we focus on constrained, nonlinear optimization. Formally, we consider the following global minimization problems:

$$\min_x f(x),$$

$$h_i(x) = 0 \quad i \in E$$

$$g_j(x) \leq 0 \quad j \in I$$

$$a \leq x \leq b,$$

where we search for to the global minimizer point of f, an n-dimensional, real function. The equality and inequality constraint functions h_i and g_j and the lower and upper bounds a and b of the n-dimensional variable vectors determine the feasible set of points. If the constraints are present, we optimize over a new objective function, denoted by F, having the same optimum points as the original function, but it represents the constraints by penalty terms. These increasingly add more and more to the value of f as the point that we evaluate F gets farther and farther from the feasible set of points (for details see [65]). This function replacement results in a much simpler problem:

$$\min_{a \leq x \leq b} F(x).$$

The violation of the upper and lower bounds can easily be avoided if candidate extremal points are generated within the allowed interval during the optimization.

1.3 The GLOBAL Algorithm

GLOBAL is a stochastic, multistart type algorithm that uses clustering to increase efficiency. Stochastic algorithms assume that the problem to be solved is not hopeless in the sense that the relative size of the region of attraction of the global minimizer points is not too small, i.e., we have a chance to find a starting point in these regions of attraction that can then be improved by a good local search method. Anyway, if you use a stochastic algorithm, then running it many times may result in different results. Also, it is important to set the sampling density with care to achieve good level of reliability while keeping speed. The theoretical framework of stochastic global optimization methods is described in respective monographs such as [63, 74]. The specific theoretical results valid for the GLOBAL algorithm are given in the papers [7, 8, 11, 59, 60]. Among others, these state that the expected number of local searches is finite with probability one even if the sampling goes for ever.

Deterministic global optimization methods have the advantage of sure answers compared to the uncertain nature of the results of stochastic techniques (cf. [27, 72]).

On the other hand, deterministic methods [66] are usually more sensitive for the dimension of the problem, while stochastic methods can cope with larger dimensional ones. The clustering as a tool for achieving efficiency in terms of the number of local searches was introduced by Aimo Törn [71].

Algorithm 1.1 GLOBAL

Input

$F: \mathbb{R}^n \to \mathbb{R}$
$a, b \in \mathbb{R}^n$: lower and upper bounds

Return value

$opt \in \mathbb{R}^n$: a global minimum candidate

1: $i \leftarrow 1, N \leftarrow 100, \lambda \leftarrow 0.5, opt \leftarrow \infty$
2: *new, unclustered, reduced, clustered* $\leftarrow \{\}$
3: **while** *stopping criteria* **is** false **do**
4:     *new* $\leftarrow$ *new* $\cup$ **generate** N **sample** from $[a, b]$ distributed *uniformly*
5:     *merged* $\leftarrow$ **sort** *clustered* $\cup$ *new* by *ascending* order regarding F
6:     *last* $\leftarrow i \cdot N \cdot \lambda$
7:     *reduced* $\leftarrow$ **select** $[0, ..., last]$ **element** from *merged*
8:     $x^* \leftarrow$ **select** $[0]$ **element** from *reduced*
9:     *opt* $\leftarrow$ **minimum** of $\{opt, x^*\}$
10:     *clustered, unclustered* $\leftarrow$ **cluster** *reduced*
11:     *new* $\leftarrow \{\}$
12:     **while size** of *unclustered* > 0 **do**
13:         $x \leftarrow$ **pop** from *unclustered*
14:         $x^* \leftarrow$ **local search** over F from x within $[a, b]$
15:         *opt* $\leftarrow$ **minimum** of $\{opt, x^*\}$
16:         **cluster** x^*
17:         **if** x^* **is** not clustered **then**
18:             **create cluster** from $\{x^*, x\}$
19:         **end if**
20:     **end while**
21:     $i \leftarrow i + 1$
22: **end while**
23: **return** *opt*

Generally, global optimization is a continuous, iterative production of possible optimizer points until some stopping condition is met. The GLOBAL algorithm creates candidate solutions in two different ways. First, it generates random samples within the given problem space. Second, it starts local searches from promising sample points that may lead to new local optima.

If multiple, different local searches lead to the same local minima, then we gained just confirmation. According to the other viewpoint, we executed unnecessary or, in other words, redundant computation. This happens when many starting points are in the same region of attraction. In this context, the region of attraction of a local minimum x^* is the set of points from which the local search will lead to x^*. With precaution, this inefficiency is reducible if we try to figure out which points are in the same region of attraction. GLOBAL achieves this through clustering.

Before any local search could take place, the algorithm executes a clustering step with the intent to provide information about the possible region of attractions of newly generated samples. Points being in the same cluster are considered as they belong to the same region of attraction. Relying on this knowledge, GLOBAL starts local searches only from unclustered samples, points which cannot be assigned to already found clusters based on the applied clustering criteria. These points might lead us to currently unknown and possibly better local optima than what we already know. The clustering is not definitive in the sense that local searches from unclustered points can lead to already found optima, and points assigned to the same cluster can actually belong to different regions of attraction. It is a heuristic procedure.

After discussing the motivation and ideas behind, let us have a look over the algorithm at the highest level of abstraction (see the pseudo code of Algorithm 1.1). There are two fundamental variables beside i: the loop counter, which are N, the sampling size, and λ, the reduction ratio that modifies the extent of how many samples are carried over into the next iteration. These are tuning parameters to control the extent of memory usage.

The random samples are generated uniformly with respect to the lower and upper bound vectors of a and b. Upon sample creation and local searching, variables are scaled to the interval $[-1, 1]^n$ to facilitate computations, and samples are scaled back to the original problem space only when we evaluate the objective function.

A modified single-linkage clustering method tailored to the special needs of the algorithm is responsible for all clustering operations. The original single-linkage clustering is an agglomerative, hierarchical concept. It starts from considering every sample a cluster on its own, then it iteratively joins the two clusters having the closest pair of elements in each round. This criterion is local; it does not take into account the overall shape and characteristics of the clusters; only the distance of their closest members matters.

The GLOBAL single-linkage interpretation follows this line of thought. An unclustered point x is added to the first cluster that has a point with a lower objective function than what x has, and it is at least as close to x as a predefined critical distance d_c determined by the formula

$$d_c = \left(1 - \alpha^{\frac{1}{N-1}}\right)^{\frac{1}{n}},$$

where n is the dimension of F and $0 < \alpha < 1$ is a parameter of the clustering procedure. The distance is measured by the infinity norm instead of the Euclidean norm. You can observe that d_c is adaptive meaning that it becomes smaller and smaller as more and more samples are generated.

The latest available version of GLOBAL applies local solvers for either differentiable or non-differentiable problems. FMINCON is a built-in routine of MATLAB using sequential quadratic programming that relies on the BFGS formula when updating the Hessian of the Lagrangian. SQLNP is a gradient-based method capable of solving linearly constrained problems using LP and SQP techniques. For non-differentiable cases, GLOBAL provides UNIRANDI, a random walk type method. We are going to discuss this stochastic algorithm in detail later along our improvements and proposals. For the line search made in UNIRANDI, one could also apply stochastic process models, also called Bayesian algorithm [40].

The original algorithm proposed by Boender et al. [11] stops if in one main iteration cycle no new local minimum point is detected. This reflects the assumption that the algorithm parameters are set in such a way that during one such main iteration cycle, the combination of sampling, local search, and clustering is capable to identify the best local minimizer points. In other words, the sample size, sample reduction degree, and critical distance for clustering must be determined in such a way that most of the local minima could be identified within one main iteration cycle. Usually, some experience is needed to give suitable algorithm parameters. The execution example details in Chapter 5 and especially those in Section 5.3 will help the interested reader. Beyond this, some general criteria of termination, like exhausting the allowed number of function evaluations, iterations, or CPU time can be set to stop if it found a given number of local optima or executed a given number of local searches.

Global can be set to terminate if any combination of the classic stopping rules holds true:

1. the maximal number of local minima reached,
2. the allowed number of local searches reached,
3. the maximal number of iterations reached,
4. the maximal number of function evaluations reached, or
5. the allowed amount of CPU time used.

The program to be introduced in the present volume can be downloaded from the address:

```
http://www.inf.u-szeged.hu/global/
```

Chapter 2
Local Search

2.1 Introduction

The GLOBAL method is characterized by a global and a local phase. It starts a local search from a well-chosen initial point, and then the returned point is saved and maintained by the GLOBAL method. Furthermore, there are no limitations regarding the features of the objective functions; hence an arbitrary local search method can be attached to GLOBAL. Basically, the local step is a completely separate module from the other parts of the algorithm. Usually, two types of local search are considered: methods which rely on derivative information and those which are based only on function evaluations. The latter group is also called direct search methods [63]. Naturally, the performance of GLOBAL on a problem depends a lot on the applied local search algorithm. As there are many local search methods, it is not an easy task to choose the proper one.

Originally [12], GLOBAL was equipped with two local search methods: a quasi-Newton procedure with the Davidon-Fletcher-Powell (DFP) update formula [19] and a direct search method called UNIRANDI [29]. The quasi-Newton local search method is suitable for problems having continuous derivatives, while the random walk type UNIRANDI is preferable for non-smooth problems.

In [15], a MATLAB version of the GLOBAL method was presented with improved local search methods. The DFP local search algorithm was replaced by the better performing BFGS (Broyden-Fletcher-Goldfarb-Shanno) variant. The UNIRANDI procedure was improved so that the search direction was selected by using normal distribution random numbers instead of uniform distribution ones. As a result, GLOBAL become more reliable and efficient than the old version.

B. Bánhelyi et al., *The GLOBAL Optimization Algorithm*, SpringerBriefs in Optimization, https://doi.org/10.1007/978-3-030-02375-1_2

GLOBAL was compared with well-known algorithms from the field of global optimization within the framework of BBOB 2009.[1] BBOB 2009 was a contest of global optimization algorithms, and its aim was to quantify and compare the performance of optimization algorithms in the COCO[2] framework. Now, GLOBAL was equipped with fminunc, a quasi-Newton local search method of MATLAB and with the Nelder-Mead [46] simplex method implemented as in [36]. GLOBAL performed well on ill-conditioned functions and on multimodal weakly structured functions. These aspects were also mentioned in [24], where GLOBAL was ranked together with NEWUOA [55] and MCS [28] as best for a function evaluation budget of up to $500n$ function values. The detailed results can be found in [50].

One of the main features of the UNIRANDI method is its reliability, although the algorithm may fail if the problem is characterized by long narrow valleys or the problem is ill-conditioned. This aspect is more pronounced as the dimension grows. Recently we investigated an improved variant of the UNIRANDI method [48, 49] and compared it with other local search algorithms.

Our aim now is to confirm the efficiency and reliability of the improved UNIRANDI method by including in the comparisons of other well-known derivative-free local search algorithms.

2.2 Local Search Algorithms

2.2.1 Derivative-Free Local Search

Derivative-free optimization is an important branch of the optimization where usually no restrictions are applied to the optimization method regarding the derivative information. Recently a growing interest can be observed to this topic from the scientific community. The reason is that many practical optimization problems can only be investigated with derivative-free algorithms. On the other hand with the increasing capacity of the computers and with the available parallelization techniques, these problems can be treated efficiently.

In this chapter, we consider derivative-free local search methods. They may belong to two main groups: direct search methods and model-based algorithms. The first group consists of methods like the simplex method [46], coordinate search [54], and pattern search, while in the second group belong trust-region type methods like NEWUOA [55]. The reader can find descriptions of most of these methods in [61].

---

[1] http://www.sigevo.org/gecco-2009/workshops.html#bbob.

[2] http://coco.gforge.inria.fr.

2.2.2 The Basic UNIRANDI Method

UNIRANDI is a very simple, random walk type local search method, originally proposed by Järvi [29] at the beginning of the 1970s and later used by A.A. Törn in [71]. UNIRANDI was used together with the DFP formula as part of a clustering global optimization algorithm proposed by Boender et al. [11]. Boender's algorithm was modified in several points by Csendes [12], and with the two local search methods (UNIRANDI and DFP), the algorithm was implemented called GLOBAL.

UNIRANDI relies only on function evaluations and hence can be applied to problems where the derivatives don't exist or they are expensive to evaluate. The method has two main components: the trial direction generation procedure and the line search step. These two steps are executed iteratively until some stopping condition is met.

Algorithm 2.1 shows the pseudocode of the basic UNIRANDI local search method. The trial point computation is based on the current starting point, on the generated random direction (d), and on the step length parameter (h). The parameter h has a key role in UNIRANDI since it's value is adaptively changed depending on the successful or unsuccessful steps. The opposite direction is tried if the best function value can't be reduced along the current direction (line 11). The value of the step length is halved if none of the two directions were successful (line 19).

A discrete line search (Algorithm 2.2) is started if the current trial point decreases the best function value. It tries to achieve further function reduction along the current direction by doubling the actual value of step length h until no more reduction can be achieved. The best point and the actual value of the step length are returned.

2.2.3 The New UNIRANDI Algorithm

One drawback of the UNIRANDI method is that it performs poorly on ill-conditioned problems. This kind of problems is characterized by long, almost parallel contour lines (see Figure 2.1); hence function reduction can only be achieved along hard-to-find directions. Due to the randomness of the UNIRANDI local search method, it is even harder to find good directions in larger dimensions. As many real-world optimization problems have this feature of ill-conditioning, it is worth to improve the UNIRANDI method to cope successfully with this type of problems.

Algorithm 2.1 The basic UNIRANDI local search method

Input

    f - the objective function
    x_0 - the starting point
    tol - the threshold value of the step length

Return value

    x_{best}, f_{best} - the best solution found and its function value

  1: $h \leftarrow 0.001$
  2: $fails \leftarrow 0$
  3: $x_{best} \leftarrow x_0$
  4: **while** convergence criterion is not satisfied **do**
  5:      $d \sim N(\mathbf{0}, \mathbf{I})$
  6:      $x_{trial} \leftarrow x_{best} + h \cdot d$
  7:      **if** $f(x_{trial}) < f(x_{best})$ **then**
  8:          $[x_{best}, f_{best}, h] \leftarrow LineSearch(f, x_{trial}, x_{best}, d, h)$
  9:          $h \leftarrow 0.5 \cdot h$
10:      **else**
11:          $d \leftarrow -d$
12:          $x_{trial} \leftarrow x_{best} + h \cdot d$
13:          **if** $f(x_{trial}) < f(x_{best})$ **then**
14:              $[x_{best}, f_{best}, h] \leftarrow LineSearch(f, x_{trial}, x_{best}, d, h)$
15:              $h \leftarrow 0.5 \cdot h$
16:          **else**
17:              $fails \leftarrow fails + 1$
18:              **if** $fails \geq 2$ **then**
19:                  $h \leftarrow 0.5 \cdot h$
20:                  $fails \leftarrow 0$
21:                  **if** $h < tol$ **then**
22:                      **return**
23:                  **end if**
24:              **end if**
25:          **end if**
26:      **end if**
27: **end while**

Coordinate search methods iteratively perform line search along one axis direction at the current point. Basically, they are solving iteratively univariate optimization problems. Well-known coordinate search algorithms are the Rosenbrock method [63], the Hooke-Jeeves algorithm [26], and Powell's conjugate directions method [54].

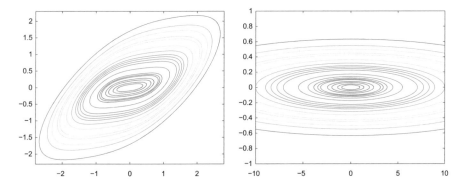

Fig. 2.1 Ill-conditioned functions

Algorithm 2.2 The *LineSearch* function

Input

    f - the objective function
    x_{trial} - the current point
    x_{best} - the actual best point
    d and h - the current direction and step length

Return value

    x_{best}, f_{best}, h - the best point, the corresponding function value, and the step length

1: **while** $f(x_{trial}) < f(x_{best})$ **do**
2:     $x_{best} \leftarrow x_{trial}$
3:     $f_{best} \leftarrow f(x_{best})$
4:     $h \leftarrow 2 \cdot h$
5:     $x_{trial} \leftarrow x_{best} + h \cdot d$
6: **end while**

The Rosenbrock method updates in each iteration an orthogonal coordinate system and makes a search along an axis of it. The Hooke-Jeeves method performs an exploratory search in the direction of coordinate axes and does a *pattern search* (Figure 2.2) in other directions. The pattern search is a larger search in the improving direction also called *pattern direction*. The Powell's method tries to discard one direction in each iteration step by replacing it with the pattern direction. One important feature of these algorithms is that they can follow easily the contour lines of the problems having narrow, turning valley-like shapes.

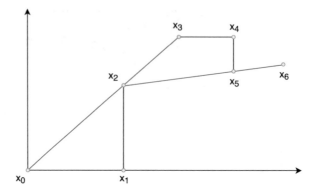

Fig. 2.2 Pattern search along the directions $x_2 - x_0$ and $x_5 - x_2$

Inspired by the previously presented coordinate search methods, we introduced some improvements to the UNIRANDI algorithm. The steps of the new method are listed in Algorithm 2.3. The UNIRANDI local search method was modified so that after a given number of successful line searches along random directions (lines 7–36), two more line searches are performed along pattern directions (lines 39–53).

Algorithm 2.3 The new UNIRANDI local search algorithm

Input

  f - the objective function
  x_0 - the starting point
  tol - the threshold value of the step length

Return value

  x_{best}, f_{best} - the best solution found and its function value

1: $h \leftarrow 0.001$
2: $fails \leftarrow 0$
3: $x_{best} \leftarrow x_0$
4: **while** convergence criterion is not satisfied **do**
5:     $itr \leftarrow 0$
6:     Let $dirs_i, i = 1 \ldots maxiters$ initialized with the null vector
7:     **while** $itr < maxiters$ **do**
8:         $d \sim N(\mathbf{0}, \mathbf{I})$
9:         $x_{trial} \leftarrow x_{best} + h \cdot d$
10:         **if** $f(x_{trial}) < f(x_{best})$ **then**
11:             $[x_{best}, f_{best}, h] \leftarrow LineSearch(f, x_{trial}, x_{best}, d, h)$
12:             $h \leftarrow 0.5 \cdot h$
13:             $fails \leftarrow 0$
14:             $itr \leftarrow itr + 1$
15:             $dirs_{itr} \leftarrow x_{best} - x_0$
16:         **else**
17:             $d \leftarrow -d$
18:             $x_{trial} \leftarrow x_{best} + h \cdot d$
19:             **if** $f(x_{trial}) < f(x_{best})$ **then**

```
20:                         [x_best, f_best, h] ← LineSearch(f, x_trial, x_best, d, h)
21:                         h ← 0.5 · h
22:                         fails ← 0
23:                         itr ← itr + 1
24:                         dirs_itr ← x_best − x_0
25:                     else
26:                         fails ← fails + 1
27:                         if fails ≥ 2 then
28:                             h ← 0.5 · h
29:                             fails ← 0
30:                             if h < tol then
31:                                 return
32:                             end if
33:                         end if
34:                     end if
35:                 end if
36:         end while
37:         The best point is saved as the starting point for the next iteration: x_0 ← x_best
38:         Let d_1 ← dirs_itr and d_2 ← dirs_{itr−1} the last two pattern directions saved during the pre-
            vious iterations
39:         for i ∈ {1, 2} do
40:             x_trial ← x_best + h · d_i
41:             if f(x_trial) < f(x_best) then
42:                 [x_best, f_best, h] ← LineSearch(f, x_trial, x_best, d_i, h)
43:                 h ← 0.5 · h
44:             else
45:                 d_i ← −d_i
46:                 x_trial ← x_best + h · d_i
47:                 if f(x_trial) < f(x_best) then
48:                     [x_best, f_best, h] ← LineSearch(f, x_trial, x_best, d_i, h)
49:                     h ← 0.5 · h
50:                 end if
51:             end if
52:         end for
53: end while
```

After each successful line search, new pattern directions are computed and saved in lines 15 and 24. The role of the $fails$ variable is to follow the unsuccessful line search steps, and the value of the h parameter is halved if two consecutive failures occur (line 27). This step prevents h to decrease quickly, hence avoiding a premature exit from the local search algorithm.

In the final part of the algorithm (lines 39–52), pattern searches are performed along the last two pattern directions ($dirs_{itr}$ and $dirs_{itr−1}$) computed during the previous iterations.

These steps can be followed in two-dimensions on Figure 2.3. After three searches along random directions (d_1, d_2, and d_3), we take two line searches along the pattern directions $x_3 − x_0$ and $x_2 − x_0$ in order to speed-up the optimization process.

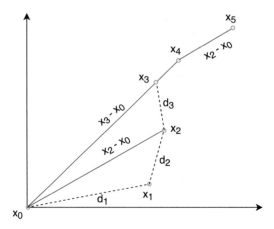

Fig. 2.3 Improved UNIRANDI: search along pattern directions $x_3 - x_0$ and $x_2 - x_0$

2.2.4 Reference Algorithms

The following derivative-free algorithms are considered for comparison in the subsequent sections: Nelder-Mead simplex method (NM), Powell's conjugate gradient method (POWELL), Hooke-Jeeves algorithm (HJ), and NEWUOA, a model-based method.

Nelder-Mead Simplex Method [46] The method uses the concept of a simplex, a set of $n+1$ points in n dimension. The algorithm evaluates all the $n+1$ points of the simplex and attempts to replace the point with the largest function value with a new candidate point. The new point is obtained by transforming the worst point around the centroid of the remaining n points using the following operations: reflection, expansion, and contraction.

As the shape of the simplex can be arbitrarily flat, it is not possible to prove global convergence to stationary points. The method can stagnate and converge to a non-stationary point. In order to prevent stagnation, Kelley [35] proposed a restarted variant of the method that uses an approximate steepest descent step.

Although the simplex method is rather old, it still belongs to the most reliable algorithms especially in lower dimensions [52]. In this study, we use the implementation from [35].

Powell's Algorithm [54] It tries to construct a set of conjugate directions by using line searches along the coordinate axes. The method initialize a set of directions u_i to the unit vectors e_i, $i = 1, \ldots, n$. A search is started from an initial point P_0 by performing n line searches along directions u_i. Let P_n be the point found after n line searches. After these steps the algorithm updates the set of directions by eliminating the first one ($u_i = u_{i+1}$, $i = 1, \ldots, n-1$) and setting the last direction to $P_n - P_0$. In the last step, one more line search is performed along the direction u_n.

In [43], a recent application of Powell's method is presented. We used a MAT-LAB implementation of the method described in [56].

Hooke-Jeeves Method [26] It is a pattern search technique which performs two types of search: an exploratory search and a pattern move. The exploratory search is a kind of a neighboring search where the current point is perturbed by small amounts in each of the variable directions. The pattern move is a longer search in the improving direction. The algorithm makes larger and larger moves as long as the improvement continues. We used the MATLAB implementation from [36].

NEWUOA Algorithm [55] NEWUOA is a relatively new local search method for unconstrained optimization problems. In many papers [18, 24, 61, 21], it appeared as a reference algorithm, and it is considered to be a state-of-the-art solver.

The algorithm employs a quadratic approximation of the objective function in the trust region. In each iteration, the quadratic model interpolates the function at $2n + 1$ points. The remaining degree of freedom is taken up by minimizing the Frobenius norm of the difference between the actual and previous model.

We used the implementation from NLopt [30] through the OPTI TOOLBOX [17] which offered a MATLAB MEX interface.

2.3 Computational Investigations

2.3.1 Experimental Settings

We have conducted some computational simulations as follows: at first, the improved UNIRANDI algorithm (nUNIR) was compared with the previous version (UNIR) in terms of reliability and efficiency. The role of the second experiment is to explore the differences between the new UNIRANDI method and the reference algorithms presented previously. In the third experiment, all the local search methods were tested in terms of error value, while in the final stage, the performance of the methods was measured in terms of percentage of solved problems. During the simulations, the local search methods were tested as a part of the GLOBAL algorithm.

The testbed consists of 63 problems with characteristics like separability, non-separability, and ill-conditioning. For some problems, the rotated and shifted versions were also considered. Thirty-eight of the problems are unimodal, and 25 are multimodal with the dimensions varying between 2 and 60.

The main comparison criteria are the following: the average number of function evaluations (NFE), the success rate (SR), and the CPU time. SR equals to the ratio of the number of successful trials to the total number of trials expressed as a percentage. A trial is considered successful if $|f^* - f_{best}| \leq 10^{-8}$ holds, where f^* is the known global minimum value, while f_{best} is the best function value obtained. The function evaluations are not counted if a trial fails to find the global minimum;

hence it counts as an unsuccessful run. The different comparison criteria are computed over 100 independent runs with different random seeds. In order to have a fair comparison, the same random seed was used with each local search algorithm.

The maximal allowed function evaluation budget during a trial was set to $2 \cdot 10^4 \cdot n$. The GLOBAL algorithm runs until it finds the global optimum with the specified precision or when the maximal number of function evaluations is reached. In each iteration of GLOBAL, 50 random points were generated randomly, and the 2 best points were selected for the reduced sample. A local search procedure inside the GLOBAL algorithm stops if it finds the global optimum with the specified precision or the relative function value is smaller than 10^{-15}. They also stop when the number of function evaluations is larger than half of the total available budget.

During the optimization process, we ignored the boundary handling technique of the UNIRANDI method since the other algorithms do not have this feature.

All computational tests have been conducted under MATLAB R2012a on a 3.50 GHz Intel Core i3 machine with 8 Gb of memory.

2.3.2 Comparison of the Two UNIRANDI Versions

In the current subsection, we analyze the reliability and the efficiency of the two versions of the UNIRANDI method. The comparison metrics were based on the average function evaluations, success rate, and CPU time. The corresponding values are listed in Tables 2.1 and 2.2.

The success rate values show that the new UNIRANDI local search method is more reliable than the old one. Usually, the earlier, called nUNIR, has larger or equal SR values than UNIR, except the three multimodal functions (Ackley, Rastrigin, and Schwefel). UNIR fails to converge on ill-conditioned functions like Cigar, Ellipsoid, and Sharpridge but also on problems that have a narrow curved valley structure like Rosenbrock, Powell, and Dixon-Price. The available budget is not enough to find the proper directions on these problems. The SR value of the nUNIR method is almost 100% in most of the cases except some hard multimodal functions like Ackley, Griewank, Perm, Rastrigin, and Schwefel.

Considering the average number of function evaluations, nUNIR requires less number of function evaluations than UNIR, especially on the difficult problems. nUNIR is much faster on the ill-conditioned problems, and the differences are more pronounced in larger dimensions (e.g., Discus, Sum Squares, and Zakharov). On the problems with many local minima, the nUNIR is again faster than UNIR except the Ackley, Griewank, and Schwefel functions. The CPU time also reflects the superiority of the nUNIR method over UNIR on most of the problems.

The last line of Tables 2.1 and 2.2 show the average values of the indicators (NFE, CPU) computed over those problems where at least one trial was successful for both of the methods. The SR is computed over the entire testbed. The aggregated values of NFE, SR, and CPU time again show the superiority of the nUNIR method over UNIR.

Table 2.1 Comparison of the two versions of the UNIRANDI method in terms of number of function evaluations (NFE), success rate (SR), and CPU time—part 1

Function	dim	UNIR			nUNIR		
		NFE	SR	CPU	NFE	SR	CPU
Ackley	5	25,620	93	0.5479	32,478	87	0.7485
Beale	2	3133	98	0.0353	3096	98	0.0356
Booth	2	168	100	0.0062	185	100	0.0067
Branin	2	172	100	0.0064	170	100	0.0064
Cigar	5	68,357	41	0.5072	542	100	0.0133
Cigar	40	–	0	4.3716	6959	100	0.1140
Cigar-rot	5	57,896	57	0.9968	930	100	0.0317
Cigar-rot	40	–	0	11.9116	16,475	100	0.4034
Cigar-rot	60	–	0	20.3957	28,062	100	0.6363
Colville	4	15,361	100	0.1399	1524	100	0.0226
Diff. Powers	5	–	0	0.7139	1926	100	0.0340
Diff. Powers	40	–	0	9.7129	91,786	100	1.2864
Diff. Powers	60	–	0	18.5559	189,939	100	3.1835
Discus	5	5582	100	0.0555	1807	100	0.0253
Discus	40	23,480	100	0.2309	19,484	100	0.2354
Discus-rot	5	5528	100	0.1342	4477	100	0.1232
Discus-rot	40	23,924	100	0.4317	20,857	100	0.4153
Discus-rot	60	30,910	100	0.5558	27,473	100	0.5196
Dixon-Price	10	74,439	80	0.6853	15,063	100	0.1850
Easom	2	717	100	0.0133	1629	100	0.0295
Ellipsoid	5	41,611	100	0.2998	976	100	0.0170
Ellipsoid	40	–	0	9.7700	44,619	100	0.7159
Ellipsoid-rot	5	41,898	100	0.6076	3719	100	0.1058
Ellipsoid-rot	40	–	0	17.1149	71,799	100	1.7026
Ellipsoid-rot	60	–	0	26.7447	120,476	100	2.9285
Goldstein Price	2	233	100	0.0064	228	100	0.0072
Griewank	5	43,749	34	0.7231	44,944	34	0.7816
Griewank	20	12,765	100	0.1839	11,801	100	0.1792
Hartman	3	878	100	0.0298	241	100	0.0128
Hartman	6	9468	100	0.2168	1056	100	0.0513
Levy	5	31,976	77	0.6050	17,578	99	0.3083
Matyas	2	172	100	0.0062	188	100	0.0069
Perm-(4,1/2)	4	–	0	0.7295	44,112	44	0.7426
Perm-(4,10)	4	5076	1	0.8043	16,917	99	0.2437
Powell	4	43,255	33	0.5409	1787	100	0.0359
Powell	24	–	0	3.5950	42,264	100	0.4767
Power Sum	4	16,931	10	0.7003	33,477	86	0.4677
Rastrigin	4	36,665	22	0.6912	34,449	21	0.7817
Average		20,746	60	0.3247	**11,405**	**95**	**0.2043**

Table 2.2 Comparison of the two versions of the UNIRANDI method in terms of number of function evaliations (NFE), success rate (SR), and CPU time—part 2

Function	dim	UNIR			nUNIR		
		NFE	SR	CPU	NFE	SR	CPU
Rosenbrock	5	–	0	0.6763	2227	100	0.0415
Rosenbrock	40	–	0	5.7691	70,624	100	0.7062
Rosenbrock-rot	5	–	0	1.5023	1925	100	0.0719
Rosenbrock-rot	40	–	0	12.7113	78,104	100	1.4457
Rosenbrock-rot	60	–	0	20.0845	137,559	100	2.3171
Schaffer	2	18,728	36	0.2469	14,270	94	0.1499
Schwefel	5	57,720	40	0.7215	58,373	37	0.7733
Shekel-5	4	1314	100	0.0488	1401	100	0.0543
Shekel-7	4	1506	100	0.0513	1646	100	0.0616
Shekel-10	4	1631	100	0.0573	1817	100	0.0658
Sharpridge	5	–	0	0.9025	961	100	0.0209
Sharpridge	40	–	0	7.0828	12,755	100	0.2337
Shubert	2	854	100	0.0206	827	100	0.0211
Six hump	2	137	100	0.0058	139	100	0.0070
Sphere	5	292	100	0.0106	331	100	0.0122
Sphere	40	2698	100	0.0682	2799	100	0.0788
Sum Squares	5	373	100	0.0134	396	100	0.0147
Sum Squares	40	21,973	100	0.3337	8205	100	0.1696
Sum Squares	60	49,435	100	0.6189	15,053	100	0.3084
Sum Squares-rot	60	47,700	100	0.8876	17,472	100	0.4365
Trid	10	5588	100	0.0964	2057	100	0.0440
Zakharov	5	427	100	0.0148	465	100	0.0151
Zakharov	40	18,784	100	0.2799	16,913	100	0.2761
Zakharov	60	41,633	100	0.4633	36,191	100	0.4407
Zakharov-rot	60	42,813	100	0.9140	37,799	100	0.8689
Average		20,746	60	0.3247	**11,405**	**95**	**0.2043**

2.3.3 Comparison with Other Algorithms

This subsection presents comparison results between the new UNIRANDI method and the reference algorithms of Nelder-Mead, POWELL, Hooke-Jeeves, and the NEWUOA method. The main indicators of comparison are the average number of function evaluations and success rate. The results are listed in Tables 2.3 and 2.4.

Considering the success rate values, we can observe the superiority of the nUNIR and NEWUOA algorithms over the other methods. The smallest values for nUNIR are shown for some hard multimodal problems like Griewank, Perm, Rastrigin, and Schwefel with 34%, 44%, 21%, and 37%, respectively. NEWOUA has almost 100% values everywhere; however, it fails on all the trials for the Different Powers (40 and 60 dimensions) and Sharpridge (40 dimension) functions. Although the POWELL and Hooke-Jeeves techniques have similar behavior, the success rate values differ often significantly. The Nelder-Mead simplex method usually fails in the trials in higher dimensions, but in small dimensions, it shows very good values.

The average number of function evaluations show that NEWOUA is very fast and outperforms the other methods on most of the problems. Another aspect is that the rotated version of some functions usually requires substantially more function evaluations (e.g., Discus, Ellipsoid in 40 dimension, and Sum Squares in 60 dimensions). NEWOUA is followed by the POWELL method which is very fast especially on separable functions (Cigar, Discus, and Ellipsoid). The Hooke-Jeeves algorithm is less efficient than the POWELL method and is more sensitive to the rotation (see the Discus, Ellipsoid, Rosenbrock, and Sum Squares functions). The Nelder-Mead algorithm provides the best results after NEWUOA in the case of some functions (Beale, Goldstein-Price, and Powell) in lower dimension. The new UNIRANDI method is slower than the best algorithm from the counterparts. On the other hand, it is not that sensitive to the rotation as the other algorithms (see Discus, Ellipsoid, and Rosenbrock in 40 dimension).

The last two rows of Tables 2.3 and 2.4 present the average values of the indicators (NFE, SR) that are computed over those problems where at least one trial was successful for each of the five methods (Average1). The last row shows the results without the Nelder-Mead algorithm (Average2). The success rate value is computed here over the entire testbed. Considering the aggregated values, the reliability of nUNIR is better than that of the other methods, and it proved to be the second most efficient algorithm after NEWUOA.

2.3.4 Error Analysis

Comparing function error values is a widely used evaluation criterion of optimization algorithms. Using this indicator the algorithms may be compared even if they do not converge. The error value during a single trial is defined as the difference of the function values of the best point found and the known global optimum. As the algorithms are tested over several trials, we usually calculate average and median error values.

The aim of the error analysis is to evaluate the average and median errors of the six local search methods over the entire testbad. For this reason, the obtained average error values for each problem were represented in a boxplot graph (Figure 2.4). The boxplot contains the lower and upper quartiles, the median, the mean values represented by big circles, and the outliers by small circles. The plots use a logarithmic scale.

Considering the average values, the nUNIR method shows the smallest interquartile range, and has the smallest number of outliers. UNIR shows a similar lower quartile as nUNIR, but the upper quartile is much larger. The boxplot shows the best results in the case of the NEWOUA, which obtained very small error values for some functions. The POWELL method has also some small error values; however, the third quartile is much larger than in the case of the NEWUOA. The Hooke-Jeeves algorithm (HJ) spans a smaller range of values than POWELL having larger

Table 2.3 Comparison of the nUNIR, NM, POWELL, HJ, and NEWUOA methods in terms of number of function evaluations (NFE) and success rate (SR)—part 1

Function	dim	nUNIR NFE	SR	NM NFE	SR	POWELL NFE	SR	HJ NFE	SR	NEWUOA NFE	SR
Ackley	5	32,478	87	–	0	–	0	–	0	1868	100
Beale	2	3096	98	416	100	3043	98	11,439	9	363	100
Booth	2	185	100	113	100	118	100	2203	100	96	100
Branin	2	170	100	112	100	127	100	2265	100	376	100
Cigar	5	542	100	413	100	121	100	906	100	105	100
Cigar	40	6959	100	–	0	634	100	6408	100	172	100
Cigar-rot	5	930	100	428	100	432	100	4854	100	213	100
Cigar-rot	40	16,475	100	–	0	13,502	100	208,889	100	487	100
Cigar-rot	60	28,062	100	–	0	19,402	100	710,014	23	645	100
Colville	4	1524	100	512	100	1616	100	10,425	100	460	100
Diff. Powers	5	1926	100	42,169	24	370	100	11,136	100	–	0
Diff. Powers	40	91,786	100	–	0	71,098	98	9919	100	–	0
Diff. Powers	60	189,939	100	–	0	104,892	99	16,229	100	–	0
Discus	5	1807	100	337	100	121	100	843	100	105	100
Discus	40	19,484	100	–	0	644	100	5536	100	170	100
Discus-rot	5	4477	100	599	100	387	100	–	0	630	100
Discus-rot	40	20,857	100	–	0	37,496	100	493,423	87	11,075	100
Discus-rot	60	27,473	100	–	0	97,780	100	938,953	12	17,355	100
Dixon-Price	10	15,063	100	54,716	99	33,015	88	8094	100	8360	100
Easom	2	1629	100	2562	84	24,160	5	9292	59	4264	100
Ellipsoid	5	976	100	421	100	121	100	1003	100	99	100
Ellipsoid	40	44,619	100	–	0	641	100	6332	100	183	100
Ellipsoid-rot	5	3719	100	499	100	384	100	44,034	89	462	100
Ellipsoid-rot	40	71,799	100	–	0	33,003	100	126,523	100	26,282	100
Ellipsoid-rot	60	120,476	100	–	0	87,515	100	252,099	100	117,035	100
Goldstein Price	2	228	100	136	100	2272	98	3733	82	361	100
Griewank	5	44,944	34	37,925	43	73,922	1	44,848	37	14,059	36
Griewank	20	11,801	100	269,929	1	70,485	60	23,251	100	925	100
Hartman	3	241	100	160	100	147	100	1132	76	183	100
Hartman	6	1056	100	778	100	20,025	91	1293	84	401	100
Levy	5	17,578	99	10,745	100	30,517	5	33,048	89	1240	100
Matyas	2	188	100	109	100	118	100	3241	100	96	100
Perm-(4,1/2)	4	44,112	44	10,063	100	15,439	11	–	0	15,697	100
Perm-(4,10)	4	16,917	99	2213	100	11,747	42	37,309	29	1697	100
Powell	4	1787	100	276	100	729	100	17,759	29	478	100
Powell	24	42,264	100	–	0	105,542	37	–	0	14,791	100
Power Sum	4	33,477	86	8809	100	25,990	8	–	0	6117	100
Rastrigin	4	34,449	21	27,936	47	3110	5	27,993	13	33,710	20
Average1		7352	**95**	14,542	53	11,770	78	12,020	68	**3245**	89
Average2		14,612	**95**	–	–	24,860	78	67,926	68	**6945**	89

Table 2.4 Comparison of the nUNIR, NM, POWELL, HJ, and NEWUOA methods in terms of number of function evaluations (NFE), and success rate (SR)—part 2

Function	dim	nUNIR NFE	nUNIR SR	NM NFE	NM SR	POWELL NFE	POWELL SR	HJ NFE	HJ SR	NEWUOA NFE	NEWUOA SR
Rosenbrock	5	2227	100	980	100	9721	99	5281	100	685	100
Rosenbrock	40	70,624	100	–	0	512,485	1	56,340	100	13,760	100
Rosenbrock-rot	5	1925	100	1182	100	7288	99	20,585	100	640	100
Rosenbrock-rot	40	78,104	100	–	0	–	0	371,371	22	14,147	100
Rosenbrock-rot	60	137,559	100	–	0	–	0	935,695	23	21,493	100
Schaffer	2	14,270	94	4712	100	13,704	16	3233	82	3213	100
Schwefel	5	58,373	37	57,721	41	50,222	1	56,191	6	31,947	47
Shekel-5	4	1401	100	1110	100	2795	100	2662	100	588	100
Shekel-7	4	1646	100	957	100	6848	99	3641	100	493	100
Shekel-10	4	1817	100	965	100	10,681	99	5093	100	496	100
Sharpridge	5	961	100	–	0	634	100	32,400	4	41,470	2
Sharpridge	40	12,755	100	–	0	9652	100	–	0	–	0
Shubert	2	827	100	562	100	9922	59	6423	69	215	100
Six hump	2	139	100	117	100	106	100	1561	100	249	100
Sphere	5	331	100	264	100	120	100	537	100	113	100
Sphere	40	2799	100	–	0	634	100	4981	100	171	100
Sum Squares	5	396	100	305	100	125	100	826	100	94	100
Sum Squares	40	8205	100	–	0	676	100	7831	100	170	100
Sum Squares	60	15,053	100	–	0	992	100	11,353	100	213	100
Sum Squares-rot	60	17,472	100	–	0	21,692	100	60,914	100	2211	100
Trid	10	2057	100	–	0	2007	100	9749	100	1801	100
Zakharov	5	465	100	281	100	580	100	1675	100	331	100
Zakharov	40	16,913	100	–	0	53,942	100	–	0	16,222	100
Zakharov	60	36,191	100	–	0	143,548	100	–	0	32,818	100
Zakharov-rot	60	37,799	100	–	0	167,043	100	–	0	36,652	100
Average1		7352	**95**	14,542	53	11,770	78	12,020	68	**3245**	89
Average2		14,612	**95**	–	–	24,860	78	67,926	68	**6945**	89

first quartile. The Nelder-Mead (NM) method performs worst in this context showing the largest degree of dispersion of data.

The median error values show a similar behavior of the algorithms as in the case of average values. Now the third quartile of Powell's method is much better than in the previous case.

The sum of average and median error values for each local search method are reported in Table 2.5. This also contains one more row of error values summarizing all functions except those for a difficult function (Schwefel). The results show again that the new UNIRANDI method is quite reliable providing similar values as NEWUOA.

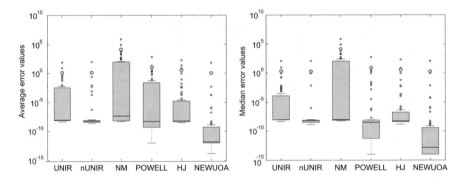

Fig. 2.4 Box plots for average (*left*) and median (*right*) errors of the local search methods

Table 2.5 Sum of average and median error values

Local search	UNIR	nUNIR	NM	POWELL	HJ	NEWUOA
Sum of averages[a]	88.55	83.93	8.5e+05	716.88	214.50	**66.84**
Sum of averages[b]	10.38	1.22	8.5e+05	55.50	35.63	**0.91**
Sum of medians[a]	129.98	**119.44**	8.5e+05	563.85	233.80	**119.44**
Sum of medians[b]	115.48	**1.00**	8.5e+05	30.87	16.66	**1.00**

[a] Sum of average/median errors over all functions
[b] Sum of average/median errors over all functions except Schwefel

2.3.5 Performance Profiles

The performance indicators computed in the previous subsections characterize the overall performance of the algorithms in an aggregate way. However, researchers may be interested in other performance indicators that reveal some different important aspects of the compared algorithms. Such performance measures were proposed in [20, 22] or more recently in [23]. In both cases, the performance of the algorithms is compared in terms of cumulative distribution functions of a given performance metric.

In our tests, we use the data profile introduced in [20] and further developed in [44]. The data profile consists of a set of problems $\mathscr{P}$, a set of solvers $\mathscr{S}$, and a convergence test. Another important ingredient of the data profile is the performance measure $t_{p,s} > 0$ for each $p \in \mathscr{P}$ problem and an $s \in \mathscr{S}$ solver. The performance measure considered in this study was the number of function evaluations.

The data profile of a solver s and α function evaluations are defined in the following way:

$$d_s(\alpha) = \frac{1}{n_p}\text{size}\{p \in \mathscr{P} : t_{p,s} \leq \alpha\}, \tag{2.1}$$

where n_p is the number of problems considered. In other words, $d_s(\alpha)$ shows the percentage of problems that can be solved with α function evaluations. Usually, there is a limit budget on the total number of function evaluations. In this study we consider larger budgets, that is, we are also interested in the long-term behavior of the examined solvers.

A problem is considered to be solved using a given solver if a convergence test is satisfied within the maximum allowed budget. The convergence test proposed in [20] is as follows:

$$f(x_0 - f(x)) \geq (1 - \tau)(f(x_0) - f_L), \qquad (2.2)$$

where $\tau \in \{10^{-1}, 10^{-3}, 10^{-5}, 10^{-7}\}$ is a tolerance parameter, x_0 is the starting point for the problem, and f_L is computed for each problem as the best value of the objective function f obtained by any solver. The convergence test measures the function value reduction obtained relative to the best possible reduction, and it is appropriate in real-world applications where the global optimum is unknown. As in our testbed, the global optimum values are known; we use the usual convergence test: $|f^* - f_{best}| \leq 10^{-8}$, where f^* is the global minimum value and f_{best} is the best function value achieved by the given solver.

We have performed several experiments by considering the different features of the test problems. Hence the whole problem set is divided into the following subsets: ill-conditioned problems, multimodal functions, low-dimensional problems (from 2 to 10 dimension), and functions with moderate or high dimensions (between 20 and 60 dimensions).

In all the tests, ten different runs of the GLOBAL method were performed on each problem with the incorporated local search algorithms. The random seeds were the same for all solvers to ensure fair comparisons. In all the scenarios, the maximum allowed function evaluations were set to 10^5. The results of data profiles for the different settings can be followed in Figures 2.5, 2.6, and 2.7. Again, all the figures use logarithmic scale.

According to Figure 2.5, NEWUOA clearly outperforms the other algorithms on the whole testbed. Although the nUNIR is slower than NEWUOA, it solves slightly more problems (85%). The POWELL and NM methods are the next fastest methods (after the NEWUOA method) until 10^3 function evaluations by solving 73% and 68% of the problems in the final stage, respectively. The Hooke-Jeeves algorithm is initially the slowest method, but in the end, it succeeded to solve 70% of the problems. UNIR is slow for budgets larger than 8000 by solving only 68% of the problems.

Considering the ill-conditioned problems (left picture of Figure 2.6), NEWUOA is again the fastest method until 20,000 function evaluations. NEWUOA is outperformed in the final stage by POWELL, nUNIR, and HJ, by solving 83%, 79%, and 74% of the problems. After a quick start, Nelder-Mead drops significantly by solving only 52% of the problems. UNIR provides the worst result (48%) on this group of problems, since the available budget usually is not enough to find a good direction.

The results on the multimodal problems (see the right picture of Figure 2.6) show different aspects compared to the previous experiments. Now after NEWUOA, NM and nUNIR are the fastest methods by solving 90% of the problems. The perfor-

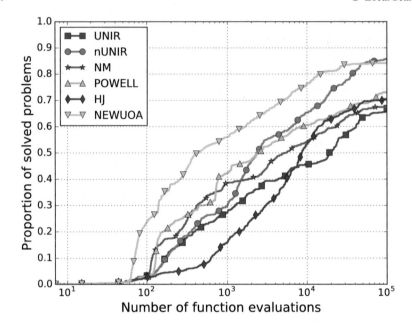

Fig. 2.5 Proportion of the solved problems over all functions

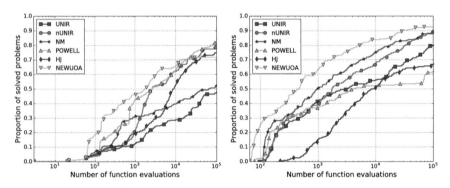

Fig. 2.6 Proportion of the solved problems for ill-conditioned (*left*) and multimodal (*right*) problems

mance of the coordinate search methods (POWELL and HJ) drops significantly by achieving a proportion of 72% and 69%, respectively. The randomness of the UNI-RANDI and the operations on simplices are more successful strategies for this group of problems.

On the low-dimensional functions (left picture of Figure 2.7), the best local solvers are the nUNIR, NEWUOA, and NM with 92%, 87%, and 85%, respectively. The coordinate search methods (POWELL and HJ) and UNIR solve around 70% of the problems. The poor performance of the POWELL and HJ methods are

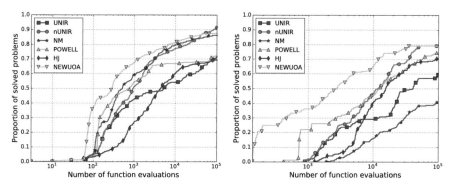

Fig. 2.7 Proportion of the solved problems for low-dimensional (*left*) and for moderate- and high-dimensional (*right*) problems

due to the multimodal problems which belong mostly to the low-dimensional set of problems. On the high-dimensional problems (right picture of Figure 2.7), the best performers are the nUNIR, NEWUOA, POWELL, and HJ by solving 80%, 79%, 75%, and 70% of the problems, respectively. Now the performance of the Nelder-Mead method drops significantly (41%) which is in accordance with the results from Tables 2.3 and 2.4.

2.4 Conclusions

Summing it up, the performance of the GLOBAL method depends much on the applied local search algorithm. The results show that both the efficiency and reliability of the new UNIRANDI method have been improved much compared to the previous variant especially on ill-conditioned problems. Compared to the other algorithms, although NEWUOA and Powell's conjugate gradient methods are usually faster, the reliability of nUNIR is promising by solving the largest number of test instances.

Chapter 3
The GLOBALJ Framework

3.1 Introduction

Henceforth in the whole chapter, we are going to use the terms *GLOBALJ*, *GLOB-ALM*, and *GLOBAL* to distinguish the concept of the new JAVA implementation, the previous MATLAB implementation [15], and the algorithm in the theoretic sense [12] in that order for the sake of clarity.

The idea, and motivation, of reworking the algorithm came while we were working on an industrial designing task [16]. GLOBAL seemed the best tool to use, but we encountered a handful of issues. First, the simulation software that calculated the score of candidate solutions could not communicate directly with any of the implementations. Second, although we were able to muster enough computing capacity, the optimization took a long time; it was barely within acceptable bounds. And last, assembling the optimization environment and fine-tuning the algorithm parameters needed much more time than it should have to. After a few weeks in the project, we decided to rewrite GLOBAL to prevent such difficulties in the future.

We planned to improve the new implementation compared to the latest one in two levels. First, we intended to create a new, easy to use and customize, modularized implementation in contrast to GLOBALM, that is highly expert-friendly, and the implementation is hard to understand. The user must be familiar with the MATLAB environment in general and with the operation of the algorithm to acquire at least some room to maneuver in customization. Second, we wanted to improve the algorithm itself, reorder the execution of base functions to achieve a better performance, and to make a parallel version of the algorithm to take advantage of multicore hardware environment. Because the parallel implementation of GLOBAL tackles interesting challenges on a different level, we dedicated the whole next chapter to its discussion, and we are going to focus exclusively on the questions of the single threaded implementation, GLOBALJ, and the algorithmic redesign.

© The Author(s), under exclusive licence to Springer International Publishing AG, part of Springer Nature 2018
B. Bánhelyi et al., *The GLOBAL Optimization Algorithm*,
SpringerBriefs in Optimization, https://doi.org/10.1007/978-3-030-02375-1_3

3.2 Switching from MATLAB to JAVA

Our first decisive action was choosing the right platform for the new implementation during the planning phase. MATLAB is a great software for many purposes; it is like a Swiss Army knife for computer science. It provides tools for almost every area of expertise ranging from image processing to machine learning. It contains the widely used algorithms and often provides more than one variants, as built-in functions, and operations are optimized for efficient handling of large data. MATLAB is great for the rapid creation of proof of concepts and research in general once someone got used to the software.

From our point of view, the first problem is its limited availability for users. MATHWORKS [69] commits a tremendous effort to maintain and improve MATLAB every year that manifests in the annual license fee; therefore, the academic institutions, universities, research groups, and larger companies are the usual users of the software. The second limitation is the result of the smaller user base, just a few software have an interface for the integration with MATLAB. Although MATLAB is shipped with an array of tools to generate code in another language, to package code as a shared library like a .NET assembly, figuring out the operation of auxiliary tools does not solve this problem, it only replaces it with another one while our original intent remains the same: we wish to compute the solution of an optimization problem. The third and last reason of change is the programming language used in MATLAB. It is an interpreted language that will always produce much slower programs than the native ones written in C++ or JAVA applications that run on a deeply optimized virtual machine.

Three programming platforms offered viable options that matched our goals, Python, C++, and JAVA. Python and JAVA are one of the most popular choices for software engineering [70], and they are used for a lot of great libraries in computer science [1, 31, 32, 47, 64, 73] and financial economics [33, 57]. C++ has lost a great portion of its fan base [70], but it is still the best language if someone wants to squeeze out the last bit of performance from his or her machine. In the end of our analysis, we decided to use JAVA for the new implementation. It provides a better performance than Python, which is also an interpreted language. It is only second to C++, but the language is familiar for much more researchers, and commercial software products often provide JAVA interfaces for integration with other applications.

3.3 Modularization

Easy customization was a priority during the design of GLOBALJ. If we look over the main optimization cycle of the algorithm in Figure 3.1, we can observe that GLOBAL can be decomposed into several key operations: generating samples, clustering samples, and executing local searches from the promising samples that probably lead to new local optima. GLOBAL organizes these operations into a pipeline of sample production and processing. The different functions are connected through the shared data that they all work on, but they are independent otherwise.

The applied local search method in GLOBAL has already been separated from the rest of the optimizer in GLOBALM, but we intended to go one step further by detaching the clustering operation too; therefore, the architecture of GLOBALJ is built up from three modules: the local search module, the clustering module, and the main module functioning as a frame for the previous two modules to implement GLOBAL. We decided to leave the sample generation in the main module because it is a very small functionality, and there is no smarter option than choosing random samples from a uniform distribution that could have motivated the creation of a fourth, sample generating module.

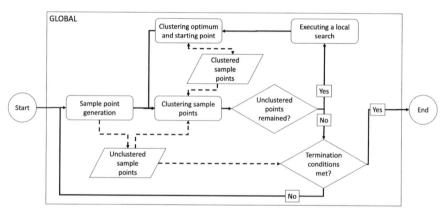

Fig. 3.1 The high-level illustration of the optimization cycle of GLOBAL. Solid lines denote control flow while dashed lines denote data flow

Besides providing the local search method for GLOBAL, the local search module is responsible for several bookkeeping operations, counting the number of executed objective function evaluations, and storing the new local optima found during the optimization. The module must be provided with the upper and lower bound of the feasible set in order to be able to restrict the searches to this region. GLOBALJ contains the implementation of the local search algorithm UNIDANDI by default.

Due to the special purpose of clustering in GLOBAL, the clustering module provides an extended interface compared to the usual operations that we would require in other contexts. This module holds the clusters, their elements, and the unclustered samples. The provided methods conform the way how GLOBAL depends on clustering; thus, there are methods for the addition, removal, and query of unclustered samples or clusters. Clustering both a single sample and all the stored, unclustered samples can be issued as well. When providing a custom implementation for GLOBALJ, we must take into account its role in the optimization. In general, the aim of running a clustering algorithm is to cluster all the samples. The algorithm only ends after every sample joined a cluster. On the contrary, the implementation of this module must be designed to execute an incomplete clustering, to identify and leave outliers alone. As GLOBAL does not use a priori knowledge about the regions of attraction, it is wise to choose a hierarchical clustering concept [45, 62] and select such a strategy which does not tend to prefer and create certain shapes of clusters, for example, spherical ones, because the real regions of attractions may have arbitrary forms. GLOBALJ contains the implementation of an improved version of the single-linkage clustering algorithm that we discuss in detail in the second part of the present chapter.

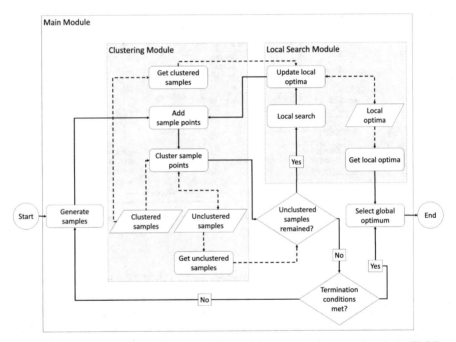

Fig. 3.2 The modules, their provided functionality, and the interaction between them in the GLOBALJ framework. As previously discussed, solid lines denote control flow while dashed lines denote data flow

The main module, being the frame algorithm of GLOBALJ, generates samples, evaluates the objective function at the samples, and moves them between the local search and clustering modules repeating the optimization cycle. Both the local search and clustering operations require the objective function values of the samples. This information was kept in a globally available data structure before, but now it is attached to each sample individually instead. We implemented the main module according to the improved algorithm of GLOBAL that will be our next topic. Figure 3.2 illustrates the cooperation of the three modules of the framework.

3.4 Algorithmic Improvements

Studying the original algorithm, we identified two points in GLOBAL that can be improved in order to avoid unnecessary local searches and replace them with much less computation expensive operations.

Algorithm 3.1 Single-linkage-clustering

Input

    F: objective-function

Input-output

    clusters: cluster-set
    unclustered: sample-set

 1: $N :=$ count(values: samples-of(*clusters*)) + size(*unclustered*)
 2: *critical-distance* := calculate-critical-distance(sample-count: N)
 3: *clustered-samples* := create-set(type: *sample*, values: *empty*)
 4: **for all** cluster: *cluster* in *clusters* **do**
 5:     **for all** sample: *sample* in *cluster* **do**
 6:         add(value: *sample*, to: *clustered-samples*)
 7:     **end for**
 8: **end for**
 9: **for all** sample: *cs* in *clustered-samples* **do**
10:     **for all** sample: *us* in *unclustered* **do**
11:         **if** distance(from: *us*, to: *cs*, type: ∞-*norm*) $\leq$ *critical-distance*
           and $F(cs) < F(us)$ **then**
12:            move(value: *us*, from: *unclustered*, to: cluster-of(*cs*))
13:         **end if**
14:     **end for**
15: **end for**
16: **return** *clusters, unclustered*

Our first point of interest was the single-linkage clustering strategy, Algorithm 3.1, applied in GLOBALM. The problem with this single-linkage interpretation realizes when a local search was executed, and its result along the starting point joined an existing cluster, or a new one was created from the two samples.

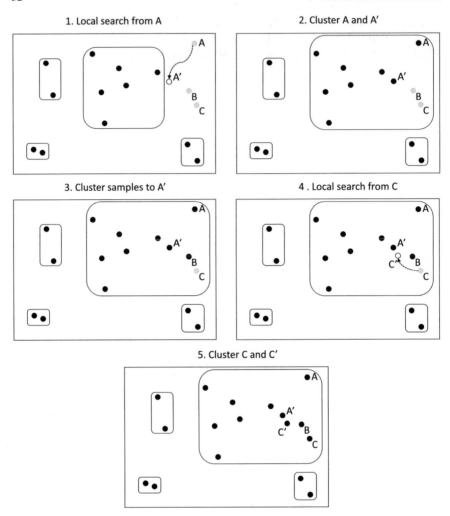

Fig. 3.3 An example scenario when the original single-linkage clustering strategy of GLOBAL fails to recognize cluster membership in time and makes an unnecessary local search as a result. Black points denote clustered samples, gray points are unclustered samples, and white points represent the result of local searches

To better understand the problem, consider the following situation that is illustrated in Figure 3.3. We have three new samples, A, B, and C, which remained unclustered after the main clustering phase of an iteration; therefore, we continue with local searches. First, we start a local search from A, and we find a cluster, the large one in the center, which has an element that is within the critical distance of A', the result point of the local search; therefore, we add A and A' to this cluster. We run a clustering step according to Algorithm 3.1 in order to look for potential cluster members within the critical distance of A and A'. As a result B also joins the center cluster. We follow the exact same process with C and add the two other sample points, C and C', to the same cluster again.

Algorithm 3.2 Recursive-single-linkage-clustering

Input

 F: objective-function

Input-output

 clusters: cluster-set
 unclustered: sample-set

 1: *newly-clustered* := create-set(type: *sample*, values: *empty*)
 2: N := count(values: samples-of(*clusters*)) + size(*unclustered*)
 3: *critical-distance* := calculate-critical-distance(sample-count: N)
 4: *clustered-samples* := create-set(type: *sample*, values: *empty*)
 5: **for all** cluster: *cluster* in *clusters* **do**
 6:   **for all** sample: *sample* in *cluster* **do**
 7:     add(value: *sample*, to: *clustered-samples*)
 8:   **end for**
 9: **end for**
10: **for all** sample: *cs* in *clustered-samples* **do**
11:   **for all** sample: *us* in *unclustered* **do**
12:     **if** distance(from: *us*, to: *cs*, type: ∞-*norm*) $\leq$ *critical-distance*
      and $F(cs) < F(us)$ **then**
13:       move(value: *us*, from: *unclustered*, to: *newly-clustered*)
14:       add(value: *us*, to: cluster-of(*cs*))
15:     **end if**
16:   **end for**
17: **end for**
18: **while** size(*newly-clustered*) > 0 **do**
19:   *buffer* := create-set(type: *sample*, values: *empty*)
20:   **for all** sample: *us* in *unclustered* **do**
21:     **for all** sample: *cs* in *newly-clustered* **do**
22:       **if** distance(from: *us*, to: *cs*, type: ∞-*norm*) $\leq$ *critical-distance*
        and $F(cs) < F(us)$ **then**
23:         move(value: *us*, from: *unclustered*, to: *buffer*)
24:         add(value: *us*, to: cluster-of(value: *cs*))
25:       **end if**
26:     **end for**
27:   **end for**
28:   *newly-clustered* := *buffer*
29: **end while**
30: **return** *clusters*, *unclustered*

The goal of clustering is to minimize the number of local searches due to their high cost. What we missed is that we could have avoided the second local search if we had realized that C is in the critical distance of B indicating that it belongs to the same cluster which A just joined; thus, the second local search was completely unnecessary. The root cause of the problem is that the algorithm goes through the samples only once in every clustering attempt and does not use immediately the new cluster information. These clustering attempts execute steps until the algorithm visited every single unclustered sample but not any further stopping in an incomplete clustering state and starting local searches instead.

We improved the single-linkage clustering of GLOBAL to filter out the above cases as well using Algorithm 3.2, an exhaustive, or recursive, clustering strategy. This means that the new clustering approach separates samples into three sets, clustered, newly clustered, and unclustered. In the beginning of each clustering attempt, only the clustered and unclustered sets have elements. If a sample fulfills the joining condition for a cluster, then it moves to the newly clustered set instead of the clustered one. After we tried to add all unclustered samples into clusters for the first time, we retry clustering the unclustered set but checking the joining condition for only the elements of the newly clustered samples. After such a follow-up attempt, newly clustered samples become clustered, and the samples added to a cluster in this iteration fill the newly clustered set. We continue these iterations until there is no movement between the sets. The recursive clustering compares each pair of samples exactly once. They do not require further objective function evaluations; moreover, a portion of these comparisons would happen anyway during the clustering steps after later local searches.

For the second time, we focused on the reduction step and cluster data handling to improve GLOBAL. The design of this algorithm and the first implementation were created when the available physical memory for programs was very limited compared to what we have today. The reduction step in the algorithm serves two purposes. It removes a given portion of the samples from the search scope, the ones having worse objective function values, as they will probably not be as good local search starting points as the rest of the samples while discarding these samples also keeps memory usage within acceptable and manageable bounds as practical algorithm design may not disregard the future execution environment. The single but significant drawback of this iterative sample reduction is that it can throw away already clustered samples as well continuously leaking the already gathered cluster information.

Figure 3.4 illustrates this undesirable effect. The portion of the sample set that have the worst objective function values is discarded in each iteration. These samples are mainly located far from the local optima that act like cluster centers from our point of view. As subsequent iterations continue, clusters become more dense because mostly the samples closer to the centers are carried over from one iteration to the other. As the critical distance decreases with the number of sample points, new clusters may appear in the place where the discarded samples of older clusters were located previously. This cluster fragmentation can be interpreted in two different ways. The first one is that GLOBAL discovers the finer granularity of the search space, and the other interpretation is that the algorithm creates false clusters in the sense that they do not represent newly discovered, separate regions of attraction that potentially mislead the search. In our experience, the latter proved to be true in the great majority of the examined cases.

1. Starting cluster

2. Sample reduction removes outer elements

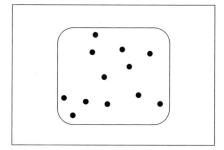

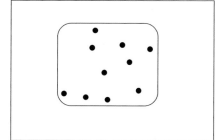

3. Shrinking continues in later iterations

4. New samples generated

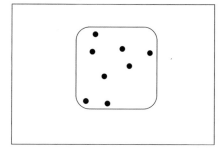

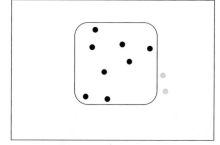

5. New cluster created from samples

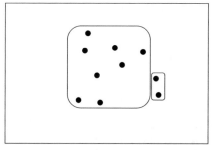

Fig. 3.4 An example of cluster erosion and fragmentation. The removal of samples with greater objective function values in the reduction step transforms clusters to be more dense and concentrated around the local minima playing the role of cluster centers in this context. This process and the decreasing critical distance over time may create multiple clusters in place of former larger ones

Cluster fragmentation raises questions about efficiency. We put significant effort into the exploration of the same part of the search space twice, or even more times. More clusters mean more local searches. GLOBAL may also finish the whole optimization prematurely reaching the maximum allowed number of local searches defined in the stopping criteria earlier. Intentionally forgetting the previously gathered cluster information can cost a lot.

The reduction step can be turned off, but we would loose its benefits as well if we decided so. Our solution is a modified reduction step that keeps the cluster membership of points, but works as before in everything else. GLOBALJ stores distinct references for the samples in all modules instead of using a single, shared, central data structure everywhere as GLOBALM does.

Algorithm 3.3 GLOBALJ

Input

> F: objective-function
> a: vector
> b: vector
> *termination*: criteria
> *clusterizer*: module
> *local-search*: module

Output

> *optimum-point*: sample
> *optimum-value*: float

1:   $optimum\text{-}value := \infty$, $optimum\text{-}point := null$, $N := 100$, $\lambda := 0.5$
2:   $search\text{-}space :=$ create-distribution(type: *uniform*, values: $[a,b]$)
3:   $reduced :=$ create-list(type: *sample*, values: *empty*)
4:   $clusters :=$ create-list(type: *cluster*, values: *empty*)
5:   $unclustered :=$ create-list(type: *sample*, values: *empty*)
6:   **while** evaluate(condition: *termination*) $= false$ **do**
7:      $new :=$ create-list(type: *sample*,
         values: generate-samples(from: *search-space*, count: N))
8:      add(values: *new*, to: *reduced*)
9:      sort(values: *reduced*, by: F, order: *descending*)
10:     remove(from: *reduced*, range: create-range(first: 1, last: $[i \cdot N \cdot \lambda]$))
11:     add(values: select(from: *reduced*, holds: in(container: *new*)),
         to: *unclustered*)
12:     $clusters, unclustered :=$ clusterizer.cluster(objective-function: F,
         unclustered: *unclustered*, clusters: *clusters*)
13:     **while** size(*unclustered*) > 0 **do**
14:        $x :=$ select(from: *unclustered*, index: 1)
15:        $x^* :=$ local-search.optimize(function: F, start: x, over: $[a,b]$)
16:        **if** $F(x^*) < optimum\text{-}value$ **then**
17:           $optimum\text{-}point := x^*$, $optimum\text{-}value := F(x^*)$
18:        **end if**
19:        $clusters, unclustered :=$ clusterizer.cluster(objective-function: F,
          unclustered: $\{x^*, x\}$, clusters: *clusters*)
20:        **if** cluster-of(x^*) $= null$ **then**
21:           $cluster :=$ create-cluster(type: *sample*, values: $\{x^*, x\}$)
22:           add(value: *cluster*, to: *clusters*)
23:        **end if**
24:        $clusters, unclustered :=$ clusterizer.cluster(objective-function: F,
          unclustered: *unclustered*, clusters: *clusters*)
25:     **end while**
26: **end while**
27: **return** *optimum-point*, *optimum-value*

As you can see in Algorithm 3.3, the pseudocode of the main module realizing GLOBAL, the reduction step still discards the references of the worst portion of samples in each iteration but does not call the removal methods of the clustering module. This approach uses more memory and spends more time in clustering than before but executes much less local searches having a reduced overall runtime in the end if we consider the whole optimization.

3.5 Results

We tested GLOBALM and GLOBALJ on a large function set to study the effects of the algorithmic improvements and the platform switch. Both implementations were provided the same memory limit and computing capacity using an average desktop PC, and we run the algorithms with the same parameter setting listed in Table 3.1. We turned off all stopping criteria except the maximum allowed number of function evaluations and the relative convergence threshold to concentrate only on the change of the number of executed function evaluations.

Table 3.1 The applied parameter values for the comparison of GLOBALJ with GLOBALM. In case of both algorithms, we ran a preliminary parameter sweep for the clustering parameter α for every function, and we used the value for which the algorithm performed the best

New samples generated in a single iteration:	400
Sample reduction factor:	5%
Maximum number of allowed function evaluations:	10^8
Relative convergence threshold:	10^{-8}
The α parameter of the critical distance:	Optimal
Applied local search algorithm:	UNIRANDI

We ran the optimizers 100 times for the entire test suite consisting of 63 frequently used functions for the performance studies of optimization methods. First, we concentrated on the executed function evaluations in our analysis. We narrowed down the study to the 25 functions for which both GLOBALJ and GLOBALM found the global optimum in all the runs in order to ensure the comparison of results of equal quality. We measured the change of the average number of executed function evaluations using the following formula:

$$change = \frac{average\ of\ GLOBALJ - average\ of\ GLOBALM}{average\ of\ GLOBALM}.$$

The results are presented in Figure 3.5. The algorithmic improvements of GLOBALJ came up to our expectation in the great majority of the cases, just as we predicted, and only performed a bit worse when it fell behind GLOBALM scoring a 27% overall improvement.

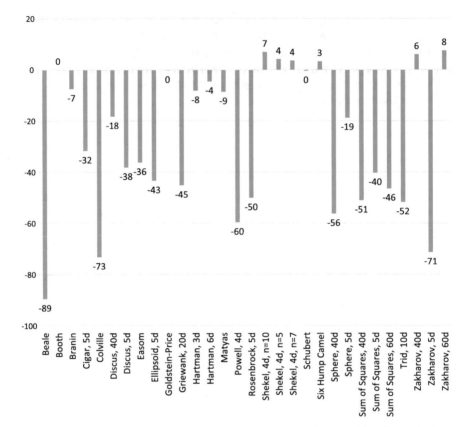

Fig. 3.5 The relative change measured in percent in the average of executed function evaluations by GLOBALJ compared to GLOBALM

GLOBALJ mainly had trouble and needed a little more effort in case of two sets of functions, the *Shekel* and the *Zakharov* functions. The former family has a lot of local optima that definitely require a higher number of local searches, and it is an exception to our general observation about cluster fragmentation. A cluster created in the early iterations of the optimization may have more than one local optima in reality in case of these functions. As an opposite, the Zakharov functions have only one global optimum each, but the search space around this point resembles to much more like a plateau. Pairing this fact with high dimensionality, running more local searches leads to the optimum earlier, and thus it is a better strategy than a decreased number of local searches.

From the technical point of view, GLOBALJ was at least ten times faster than GLOBALM in terms of runtime due to the efficiency difference of compiled and interpreted languages.

3.6 Conclusions

This chapter focused on the core structure of GLOBAL and presented our work of making the algorithm anew. We highlighted the key points of possible improvements and introduced a solution in form of a new clustering strategy. We kept the basic approach of single-linkage clustering but modified it to incorporate every available information about the search space as soon as possible and to keep all the clustering information during the whole run in order to prevent the redundant discovery of the search space. We implemented the modified GLOBAL algorithm using a modularized structure in the JAVA language to provide the key options of customization regarding the applied local solver and clustering algorithm. We compared the new optimizer and the old MATLAB implementation, and we experienced a significant improvement in the necessary function evaluations to find the global optimum.

Chapter 4
Parallelization

4.1 Introduction

The implementation of the parallel framework can operate in two different ways. The parallel structure means those units which are copyable, and their instances, may run independently from each other simultaneously. The serialized structure denotes singleton units in the optimization whose inner, lower-level operations run parallel.

Adapting to the multicore architecture of desktop and supercomputers, it seemed promising to create a parallel implementation of the Global algorithm, as we will be able to solve more difficult problems this way in reasonable time. For one hand, difficulty in our context means computationally expensive objective functions, whose evaluation can take several hours or even days for a single processor core. On the other hand, the computational complexity of a problem may come from the size of the search space that can require a lot of time to discover as well even in case of simple objective functions.

Multithreading programs make the execution faster by converting operation time to computational capacity. This conversion is 100% efficient ideally, but a lot of factors can hinder this unfortunately. Information sharing between parallel program segments is inevitable for distributing tasks and collecting results that require shared data storage that limits the number of simultaneously operating program units. Moreover, the characteristics of data flow can also degrade the efficiency. This affects the optimizer too as it disturbs the iterative algorithms the most. The algorithm depends on result of previous iterations by definition; thus they must be executed in a specific order. Although we cannot violate the principle of causality, we can still execute parallel tasks with lower efficiency in such environments.

Henceforth, we refer to the parallel version of the algorithm Global as PGlobal. It is worth to choose an environment for the implementation of PGlobal that supports fast execution and instantiation of parallel structures and possibly extends a former Global implementation. Only GlobalJ fulfills these requirements. Using the advan-

B. Bánhelyi et al., *The GLOBAL Optimization Algorithm*, SpringerBriefs in Optimization, https://doi.org/10.1007/978-3-030-02375-1_4

tages of the programming language JAVA and following a well-planned architecture, we can easily extend GlobalJ with the additional functionality.

We set several goals that the implementation of PGlobal must achieve. First, it must fit into the GlobalJ framework inheriting every key feature of the base algorithm. Second, all the previously implemented, local optimizers must be integrated into the new parallel architecture without changing their logic. Last but not least, PGlobal must have an improved performance in case of both large search space problems and high-cost objective functions. An optimizer that complies all the above requirements will be a versatile tool that is capable of handling a large variety of optimization problems with success offering customization and scalability.

4.2 Parallel Techniques

The concept of threading in data processing appeared as soon as processors started to have more than one core. Multithreading may seem to be the golden hammer for these problems, but it can bring very little or almost no improvement to the table depending on the problem. We can divide the addition of thousands of elements to sub additions, but the iterative nature of a physical simulation prevents most forms of parallelization. In the latter case, we loose the capacity surplus as resources remain idle in most of the time due to the dependencies between tasks. We can also improve iterative algorithms. For example, we can modify a random walk, local search method to create multiple starting points in each iteration executing the original single-threaded algorithm for the independent points. We select the best result after each thread finished.

It can also happen that we generate a temporary information loss, despite all the data available, that leads to reduced performance again. In this context information loss is a situation when we have every information we need, but we cannot process them in time due to parallelization resulting in the execution of unnecessary calculations.

4.2.1 Principles of Parallel Computation

We based PGlobal on the algorithm used in GlobalJ with the modified clustering strategy that we discussed in the previous chapter. You can already identify the tasks in this algorithm that differ on their input data. Such tasks can run independently as no data sharing is required between them. This is the serialized way of parallelization of the Global algorithm. This means that we simultaneously execute the disjoint tasks that were run sequentially before. This process is called functional decomposition. You can see on Figure 4.1 that the tasks A, B, and C can be executed in parallel provided they are independent from each other.

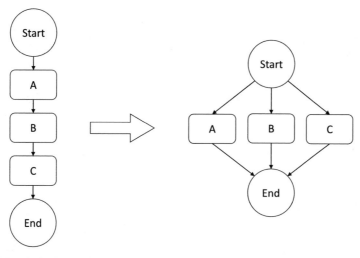

Fig. 4.1 Parallelization by functional decomposition

Another way to make a computation parallel is the simultaneous execution of the same algorithm but on multiple data packets. This type of parallelization is called data decomposition. Again, any produced data is only used within the producer execution unit. Matrix addition is a good example for this in which the same algorithm is repeated a lot of times. Data decomposition is a basis of multithreading for graphic cards. Figure 4.2 illustrates that algorithm A could work simultaneously on data 1, 2, and 3.

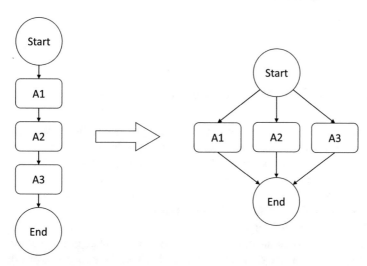

Fig. 4.2 Parallelization by data decomposition

Sometimes specific parts of an algorithm could run in parallel, while other must run sequentially due to data dependencies. The best option is the application of a pipeline structure for such systems where every part of the algorithm can work without pause. The data packets are processed by going through all the units that are responsible for the necessary data manipulations. This type of operation is considered sequential from the data point of view, while it is parallel algorithmically from the subtask point of view. Figure 4.3 shows a pipeline that executes operations A, B, and C in that order on data 1 and 2. The dotted lines denote the data flow.

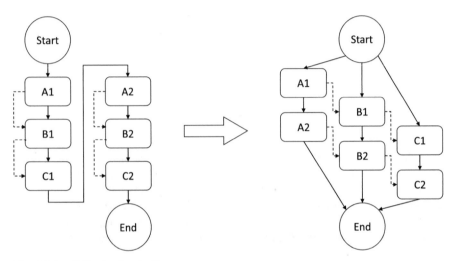

Fig. 4.3 Parallelization by pipeline structure

4.3 Design of PGLOBAL Based on GLOBAL

We developed PGlobal to combine the advantages of both the algorithm Global and the multithreaded execution. The existing algorithmic components have been modified for the parallel execution. They must be prepared so that PGlobal will not have a fixed order of execution in contrast to the sequential execution of Global. We have to organize the data flow between the computation units and the effective distribution of tasks between the different threads.

While the improved Global algorithm of GlobalJ takes advantage of sequential execution as much as possible, PGlobal runs a less efficient algorithm but uses much more computation capacity. This is necessary because the efficient sequential operation of the algorithm of GlobalJ comes from a strict order execution. Fortunately, the increased computational performance well compensates the loss of algorithmic efficiency.

PGlobal combines the advantages of functional decomposition, data decomposition, and the pipeline architecture. It implements a priority queue that distributes the tasks between the workers implementing the functional and data decomposition as multiple logical units and multiple instances of the same logical unit run simultaneously. From the data point of view, the algorithm is sequential, working like a pipeline, as each data packet goes through a fixed sequence of operations.

Figure 4.4 illustrates how the workers choose the tasks for themselves. Workers always choose the data packet that already went through the most operations. In this example, the data has to be processed by an algorithm in the stages A, B, and C of the pipeline in that order. Stage B may only work on two data packets that already went through stage A, while stage C may only work on data packets that previously processed stage B. The workers operate simultaneously and process all four data packets completely under ideal circumstances three times faster than a single-threaded algorithm would do.

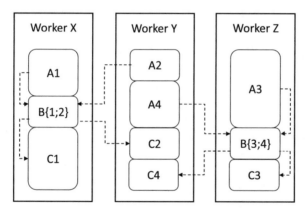

Fig. 4.4 The combination of functional decomposition, data decomposition, and the pipeline architecture

We implemented PGlobal based on the principles previously presented in this chapter. As we have already discussed, the algorithm can be separated into several different tasks. The main operations are the sample generation, the clustering, and the local searches. We study the available options of parallel computation in Global from the perspective of the data that the different logical units use and produce considering their relations and sequential dependency as well.

It is easy to see that sample generation is an independent functionality due to only depending on one static resource, the objective function. Therefore no interference is possible with any other logical component of the algorithm. Sample generation precedes the clustering in the pipeline from the data point of view.

The clustering and the local search module have a close cooperation in the improved algorithm. They work together frequently on the unclustered samples and output of resulting local searches. Scheduling the alternating repetition of these two operation types is easy in case of sequential execution. The immediate feedback from local searches at the appropriate moments provides the most complete information for clustering all the time. The improved Global uses this strategy (Figure 4.5).

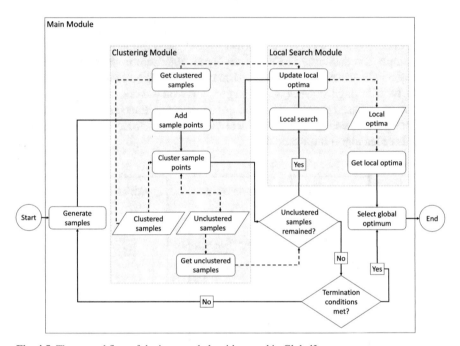

Fig. 4.5 The control flow of the improved algorithm used in GlobalJ

If we have to handle large sample sets, clustering requires a lot of resources that can only mean the usage of additional threads; therefore the parallelization of the clustering module is inevitable.

The local search during the clustering could be implemented in PGlobal by using a single thread for the searches while the others are waiting. However this part of the algorithm cannot remain sequential in the multithreaded case. Clustering becomes a quite resource heavy operation when a lot of sample points have to be handled that we must address by running the clustering operation on more threads. Using a single thread, the local search can block the execution even in the case of simple objective functions or make it run for an unacceptably long time period in case of complex objective functions. We cannot keep the improved clustering strategy and run this part of the algorithm efficiently on multiple threads at the same time; we have to choose. Fortunately with the slight modification of the improved algorithm, we are able to process a significant portion of data in parallel while we only sacrifice a little algorithmic efficiency.

We modified the algorithm in order to cease the circular dependency between the clusterizer and the local search method. We need to keep all data flow paths as we have to execute the whole algorithm on all the data, but after analyzing the problem, we can realize that not all paths have to exist continuously or simultaneously. We do not modify the algorithm if the data path going from the clusterizer to the local search method is implemented temporary. The data can wait at an execution point without data loss; therefore we can create an almost equivalent version of the improved algorithm of GlobalJ. The resulting algorithm can easily be made multithreaded using the separated modules. As long as there are samples to cluster, the program works as expected, and it integrates the found local optima in the meantime. This type of operation stops as soon as we run out of samples. This is the trigger point that activates the previously inactive, temporary data flow. To make this blocking operation as short-lived as possible, we only uphold this blocking connection while the unclustered samples are transferred from the clusterizer to the local search module. We reestablish the multithreaded operation right after this data transfer ends.

This modification results in an algorithm that moves samples in blocks into the local search module and clusters local optima immediately. This algorithm combines the improvements introduced in GlobalJ with the multithreaded execution (Figure 4.6).

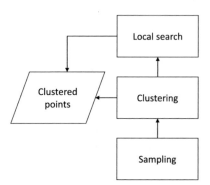

Fig. 4.6 Data paths between the algorithm components of PGlobal

4.4 Implementation of the PGlobal Algorithm

The algorithm has three logical parts. The first is the local search module, which already appears in the implementation of GlobalJ, and we have to support it. The second is the clustering module that is also a module in GlobalJ, but these modules only share the names and run different algorithms. The third is the main module which runs the PGlobal algorithm and uses the sample generation, local search, and clustering modules. The implementation of the main module is the SerializedGlobal.

SerializedGlobal and SeralizedClusterizer together are responsible for the management and preparation of multithreaded execution. The former is the optimization interface provided to the user; the latter is ready to handle the threads that enter the module and cooperates with SerializedGlobal. The operation of the two modules is closely linked.

4.4.1 SerializedGlobal

Every worker thread executes the same algorithm, PGlobal, that we illustrated on Algorithm 4.1. A thread selects a task in each main optimization cycle. We defined five different tasks; each one of them represents a data processing operation or a key part of algorithm control. Threads always choose the task that represents a latter stage in the algorithmic pipeline in order to facilitate the continuous flow of samples.

We discuss the algorithm in execution order; the tasks appear in reverse of the data flow. Each cycle starts with the check of stopping criterion.

The last stage is local search from the data point of view. The threads take samples for starting points from the *origins* synchronized queue. The thread executes the local optimization and then clusters the found local optimum and the starting point with the *clusterize optimum* algorithm (Algorithm 4.4). Threads return to the main cycle after finishing the process.

The second task is related to the termination of the algorithm. We consider the end of each clustering as the end of an iteration cycle. As soon as the number of iterations reaches the allowed maximum, and we cannot run further local searches, then we stop any sample transfer to the local search module, and the optimization ends for the worker thread.

The third task is the clustering of the samples. If the clusterizer's state is active, the thread can start the execution. Entering the clustering module is a critical phase of the algorithm, and it is guarded by a mutex synchronization primitive accordingly. After finishing the clustering, threads try to leave the module as soon as possible, while new threads may not enter. A portion of the unclustered samples is moved into the *origins* queue after all threads stopped the execution of the clustering algorithm.

The previously inactive data path is temporarily activated at this point until these samples are transferred from one module to another.

If enough samples are available, threads can choose the fourth option, setup the clustering. If the conditions are fulfilled, the samples are moved to the clusterizer, and the module state becomes active.

The last option is the sample generation task. If the number of generated samples reaches its limit, the thread stops the execution. Otherwise the threads create a single sample and store it in the *samples* shared container and check if it is a new global optimum.

With the execution of the tasks sooner or later, the system will reach a point where at least one stopping criteria is fulfilled. The main thread organizes and displays the results after all threads are terminated.

The single-threaded algorithm does not have to put too much effort to maintain consistency and keep the right order of data processing operations. The sequential execution ensures the deterministic processing of data. Meanwhile, coordinating the threads requires a lot of effort in the multithreaded case. The algorithm part responsible for keeping the consistency is in size comparable with the actual processing of data. The single thread executing Global is completely aware of the system state during the whole optimization. On the other hand, PGlobal does not have a dedicated thread for this; the system organizes itself through shared variables.

Although the implemented algorithms are totally different, the modules that handle them work similarly. The generation of a single sample happens the same way, and the only difference is thread handling in case of multiple samples. A simple cyclic structure is sufficient for the single-threaded execution that executes a predefined number of iterations. In multithreaded environment the threads must communicate with each other to achieve the same behavior. We use a shared counter whom a mutex guards. Threads may generate a new sample if and only if this counter did not reach its allowed maximum. The number of generated samples is considered to be a soft termination criterion. The algorithm can exceed the allowed maximum by a value proportional to the number of worker threads.

Local search itself works also identically; again, its controlling changed on the higher level. PGlobal uses the same local search implementations as GlobalJ with a little modification for the sake of compatibility with SerializedGlobal that completely preserves backward compatibility. The samples that can be the starting points of local searches are stored in the *origins* shared container in the parallel version of the algorithm. Threads take samples from this container and run the adapted local search algorithms started from them. After the local searches, threads finish the data processing using the *clusterize optimum* algorithm (Algorithm 4.4) as the last operation.

Algorithm 4.1 PGLOBAL

Input

    $F: \mathbb{R}^n \to \mathbb{R}$
    $a, b \in \mathbb{R}^n$: lower and upper bounds
    N: number of worker threads
    maxSampleSize: maximum number of generated samples
    newSampleSize: number of samples generated for every iteration
    reducedSampleSize: number of best samples chosen from the new samples
    batch size: number of samples forwarded to local search after clustering (if 0 use the number of currently free threads)

Return value

    optimum: best local optimum point found

```
 1: samples, unclustered, origins ← {}
 2: optimum ← maximum value
 3: start N − 1 new threads
 4: while true do
 5:        if check stopping criteria then
 6:               break
 7:        else if origins is not empty then
 8:               origin ← remove from origins
 9:               if origin is null then
10:                      continue
11:               end if
12:               localopt ← local search over F from origin within [a, b]
13:               optimum ← minimum of {optimum, localopt}
14:               call clusterize optimum (origin, localopt)
15:        else if check iteration count stopping criteria then
16:               break
17:        else if clusterizer is active then
18:               call clustering samples (critical distance)
19:               if this is last clustering thread then
20:                      origins ← remove batch size from unclustered
21:                      if |unclustered| = 0 then
22:                             set clusterizer to inactive
23:                             increase iteration count
24:                      end if
25:               end if
26:               wait until all clustering threads reach this point
27:        else if clusterizer is inactive and |samples| ≥ newSampleSize then
28:               lock samples
29:               samples ← sort samples by ascending order regarding F
30:               unclustered ← remove [1, ..., reducedSampleSize] element from samples
31:               update critical distance
32:               set clusterizer to active
33:               unlock samples
34:        else if check sample count stopping criteria then
35:               break
36:        else
37:               lock samples
38:               samples ← samples ∪ generate a new sample from [a, b] distributed uniformly
39:               optimum ← minimum of {optimum, new sample}
40:               unlock samples
41:        end if
42: end while
43: return
```

4.4.2 SerializedClusterizer

The new clustering module significantly differs from the previous implementation. It is responsible for the same work on the lower level, but it requires a much more sophisticated coordination due to the multithreading. The parallel version must also pay continuous attention to uphold consistency, while GlobalJ simply needs to iterate over the samples. The new clustering module is much more connected to the controller module. Tabs must be kept on the number of threads that is currently clustering and ones that are waiting in sleep for the end of clustering. Serialized-Global has the responsibility of managing these numbers through shared variables accessible to every thread. The manipulation of other shared variables is based on temporary privileges given exclusively to a single thread at a time. This is important for the optimal usage of the processor cores and separation of iterations. By their help and the close cooperation with the clusterizer of SerializedGlobal, we can minimize the number of local searches and the runtime.

We had to completely redesign the control flow of the clustering module as it is a crucial part from the parallelization point of view, and the previous version does not support parallel execution at all. The new module has to implement two functions. We need an ordinary clustering procedure, which assigns sample points to already existing clusters, and a secondary clustering procedure, which clusters the local optima.

Figure 4.7 shows two distinct control graphs. Although they are separated, they still share and use the same data containers whose accessibility is managed by a mutex. The dashed lines assign the mutually exclusive program parts to the lock. We implemented the clustering module in a way that makes the independent execution of these two processes possible as long as no interaction is required between them. Storing the samples and local optima in clusters and creating a new cluster are considered such interactions.

The shared variables are updated so that these types of events can keep the inner state constantly consistent. When the clusterizer is active, the worker threads continuously examine the unclustered samples. The mutual exclusion allows us to add a new cluster to the system anytime that will participate in the clustering right after its addition. The only exception to this is the case when a worker thread concludes that the clustering cannot be continued while a local optima is clustered. The thread signals the event of finished clustering by setting a flag. The termination of clustering will not be stopped despite of the new cluster; however, this does not cause any trouble in practice due to its very low frequency and insignificant consequences. Now let us discuss the *clustering samples* algorithm (Algorithm 4.2) in detail.

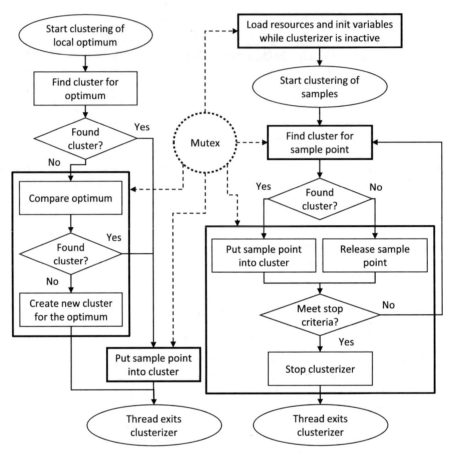

Fig. 4.7 The new clusterizer logic in PGlobal. The continuous lines denote control flow, and the dashed lines denote mutual exclusion

The operation of clusterizer depends on its inner state hidden from the outside when the clusterizer is active. The critical distance acts as an input parameter in the case of sequential execution; clustering of sample points does not affect it. However it is possible that the critical distance decreases in multithreaded environment when an optimum point is being clustered; therefore it must be handled as part of the clusterizer's state. The inner state includes the sets of unclustered and clustered samples. Clustering a local optimum affects the latter set too. The inner state is changed when the clustering ends; a portion of the unclustered samples are moved to the clustered set.

Clustering the local optima also includes the starting point of the originating search. The inner states of the clusterizer are involved again. The set of clusters might change, but the set of clustered samples and the critical distance will definitely be updated.

Algorithm 4.2 Clustering samples

Input

    critical distance: single linkage distance threshold

State before

    clustered: previously clustered samples
    unclustered: new samples to be clustered

State after

    clustered: *clustered* $\cup$ new clustered
    unclustered: *unclustered* $\setminus$ *clustered*

1: **while** clusterizer **is** active **do**
2:     *sample* $\leftarrow$ **remove** from *unclustered*
3:     **if** *sample* **is** null **then**
4:         **return**
5:     **end if**
6:     **if** *sample* **is** fully examined **then**
7:         *sample* $\rightarrow$ **insert** into *unclustered*
8:         **continue**
9:     **end if**
10:     *insider* $\leftarrow$ **find** next element in *clustered* which is not compared to sample
11:     *cluster* $\leftarrow$ null
12:     **while** *insider* **is** not null **do**
13:         **if** $\|sample - insider\|_2 \leq critical\ distance$ **and** *sample value* $>$ *insider value* **then**
14:             *cluster* $\leftarrow$ cluster of *insider*
15:             **break**
16:         **else**
17:             *insider* $\leftarrow$ **get** next element from *clustered* which is not compared to sample
18:         **end if**
19:     **end while**
20:     **lock** all cluster modifications
21:     **if** *cluster* **is** null **then**
22:         *sample* $\rightarrow$ **insert** into *unclustered*
23:     **else**
24:         *sample, cluster* $\rightarrow$ **insert** into *clustered*
25:         **if** center point of *cluster* $<$ *sample* **then**
26:             center point of *cluster* $\leftarrow$ *sample*
27:         **end if**
28:     **end if**
29:     **if** all samples are examined **then**
30:         **set** clusterizer to clean up
31:     **end if**
32:     **unlock** all cluster modifications
33: **end while**
34: **return**

The clusterizer has a variable to keep count the current clustering phase. The starting phase is the inactive one, meaning that the clusterizer is waiting for new

---

Algorithm 4.3 Compare optimum to clusters

Input

   optimum: clusterizable optimum point
   clusters: previously created clusters
   critical distance: single linkage distance threshold

Return value

   cluster: the cluster which contains the optimum

1: *cluster* ← **find** next element in *clusters* which is not compared to optimum
2: **while** *cluster* **is** not null **do**
3:       *center* ← center point of *cluster*
4:       **if** $\|center - optimum\|_2 \leq critical\ distance/10$ **then**
5:             **return** *cluster*
6:       **end if**
7:       *cluster* ← **find** next element in *clusters* which is not compared to optimum
8: **end while**
9: **return** null

---

samples to cluster. At this point only local optima can be clustered. The following phase is the active one triggered by the addition of samples denoting that clustering is in progress and further samples cannot be added. The last phase is the cleanup when the remaining unclustered samples are transferred to the local search module. The phase will become active again when a subset of unclustered samples is moved but not all. If all unclustered samples moved, the phase will be inactive again.

Considering a single-threaded execution, PGlobal works identical to GlobalM if the maximal block size is chosen; adaptive block size or block size of 1 results in equivalent algorithm to GlobalJ. Both algorithms generate samples, select a portion of them for clustering, and try to cluster these samples. GlobalJ and correctly parameterized PGlobal move exactly one sample into the local search module that another clustering attempt follows again. These steps repeat while there are unclustered samples. After evaluating every sample, a new, main optimization cycle starts in Global, and a new iteration starts in PGlobal.

When the PGlobal algorithm operates with multiple threads, it differs from GlobalJ. The clusterizer closely cooperates with the local search method in the improved Global to avoid as much unnecessary local search as possible. It makes out the most from the available data by clustering samples whenever there is an opportunity. The two modules work in turn based on the remaining unclustered sample points and compare those that were not examined yet. GlobalJ starts a local search form an unclustered sample point in case the previous clustering was incomplete. It repeats this procedure until all samples joined a cluster. The parallel version supports this activity with only moving whole sample blocks. A given amount of samples are transferred to the local search module after the clustering finished.

The maximum of transferable samples can be parameterized. The disadvantages of block movement are the possible unnecessary local searches if we transfer more samples for local search than the number of available threads. Therefore the default

---

Algorithm 4.4 Clusterize optimum

Input

    origin: starting point of the local search which lead to optimum
    optimum: optimum point to be clustered

State before

    clusters: previously created clusters
    clustered: previously clustered samples
    critical distance: single linkage distance threshold

State after

    clusters: *clusters* $\cup$ new clusters
    clustered: *clustered* $\cup$ {*origin, optimum*}
    critical distance: updated single linkage distance threshold

1: *cluster* $\leftarrow$ **call** compare optimum to clusters (*optimum, clusters, critical distance*)
2: **lock** all cluster modifications
3: **if** *cluster* **is** null **then**
4:       *cluster* $\leftarrow$ **call** compare optimum to clusters (*optimum, clusters, critical distance*)
5:       **if** *cluster* **is** null **then**
6:            *cluster* $\leftarrow$ **new** cluster
7:            center point of *cluster* $\leftarrow$ *optimum*
8:            *cluster* $\rightarrow$ **insert** into *clusters*
9:       **end if**
10: **end if**
11: *origin, cluster* $\rightarrow$ **insert** into *clustered*
12: *optimum, cluster* $\rightarrow$ **insert** into *clustered*
13: **if** center point of *cluster* $<$ *sample* **then**
14:       center point of *cluster* $\leftarrow$ *origin*
15: **end if**
16: **update** critical distance
17: **unlock** all cluster modifications

---

setting for the block size parameter is determined adaptively, and it is set to the number of threads that is currently exiting the clusterizer for optimal operation. This can be interpreted as the multithreaded extension of the single-threaded operation as it generates as much work as we are possibly able to handle. If a thread runs the local search longer than the other ones, the faster threads will automatically start clustering the search results and new local searches from the remaining samples after that. The point of this method is to keep as many samples under local search as the number of available worker threads. Balancing this way the clustering and local searching operations, we can achieve similar efficiency as the single-threaded versions do.

4.5 Parallelized Local Search

The local search methods are essential parts of the GlobalJ framework. We have to make them compatible with PGlobal to offer the same functionality. These modifications must keep backward compatibility of the resulting implementations with GlobalJ.

The parallel execution makes two requirements for the units running in multiple instances. It comes naturally that we have to create multiple algorithm instances that work in total separation from each other only sharing their source code. We have to provide an interface that facilitates the parametrization and cloning of ready-to-run instances. These clones must operate on completely different memory segments; they can only use the shared parametrization.

From a well-parameterized algorithm instance, we can easily create clones that can be run in parallel if the above two conditions are met.

4.6 Losses Caused by Parallelization

Of course, the parallel operation comes with information loss in case of PGlobal too. It is possible that the samples moved in into the local search module at the same block could be clustered using the result of a search started from this block. This can lead to unnecessary local searches; thus the program makes more computation than required degrading the efficiency. This surplus effort will be minimal if the size of the search space is large compared to the number of generated samples. It is due to the very low probability of two samples being in the same block, within the actual critical distance. This will not be a problem if the samples are generated dense. Regardless of how difficult it is to calculate the objective function, the total runtime spent on unnecessary local searches will be insignificant unless the number of executed iterations is extremely low. If we deal with less iterations, it is worth to distribute the same amount of samples between more iterations. In other words, we should increase the maximum number of iterations.

Overrunning the soft criteria of algorithm termination can lead to further losses. We can observe this phenomenon in the single-threaded versions, but parallelization creates additional opportunities for it.

4.7 Algorithm Parameters

The parametrization of the SerializedGlobal algorithm looks very similar to the parametrization of GlobalJ with a few exceptions. It is mandatory to provide the objective function and the lower and upper bounds having the same equivalent interpretation as before. The allowed maximum of the number of function evaluations, iterations, generated samples during the whole optimization, local optima, and generated samples in a single iteration are all optional. The number of worker threads and the size of sample blocks transferred from the clustering module to the local search module are additional, optional parameters. The usage of the latter is not straightforward. Positive values denote the maximal transferable amount of samples after all threads left the clustering module. All unclustered samples will certainly be moved out from the clustering module and will be used as starting points for local searches if this parameter value exceeds the number of samples generated in a single iteration. This operation is analogous to implementations prior to GlobalJ that worked the same way but on a single thread. If this parameter value is below the number of unclustered samples remained in the clusterizer, the next iteration will not start, and the idle threads start local searches. After a thread finished a local search, and there is no other sample in the local search module, it reenters the clusterizer with the result of the searches and continues the clustering of the remaining unclustered samples. It is important to note that the above-discussed sample movement can easily happen right before another thread would enter the clusterizer. This particular thread will not find any new samples, and therefore it will leave the clustering module too. This leads to the arrival of another block of new samples in the local search module eventually over feeding it. We can force a special operation if we set the size of sample blocks transferred to the local search module to the value 0. The threads leaving the clusterizer will join in all cases; thus the thread leaving last may determine the exact number of free threads. According to this parameter setting, it recalculates the maximal number of transferable samples to fit the number of threads. This results in optimal processor usage as we start precisely as many local searches as it is necessary to use all the available computing resources. Moreover, it prevents the overfeeding effect by allowing the transfer of only one sample when the latecomer threads exit.

4.8 Results

We studied the implementation of PGlobal from two aspects. First, we verified that the parallelization beneficially affects the runtime, meaning that the addition of further threads actually results in a speedup. Second, we executed a benchmark test to compare Global with PGlobal.

4.8.1 Environment

We used the configuration below for all tests.

- **Architecture:** x86_64
- **CPU(s):** 24
- **On-line CPU(s) list:** 0–23
- **Thread(s) per core:** 2

- **Core(s) per socket:** 6
- **Socket(s):** 2
- **Vendor ID:** GenuineIntel
- **CPU family:** 6
- **Model:** 44
- **Stepping:** 2
- **CPU MHz:** 1600.000
- **Virtualization:** VT-x
- **L1d cache:** 32K
- **L1i cache:** 32K
- **L2 cache:** 256K
- **L3 cache:** 12288K
- **Javac version:** 1.8.0_45
- **Java version:** 1.8.0_71

4.8.2 SerializedGlobal Parallelization Test

Parallelization tests measure the decrease of runtime on problems that are numerically equivalent in difficulty but differ in computation time. In order to make these measurements, we introduced the parameter *hardness* that affects the execution time of functions. We calculate the objective function $10^{hardness}$ times whenever we evaluate it. Thus the numerical result remains the same, but the time of function evaluation multiplies approximately by the powers of 10. The execution time greatly depends on how the system handles the hotspots. The JAVA virtual machine optimizes the code segments that make up significant portions of the execution time; thus the ten times slowdown will not happen. Considering a given hardness, we can only measure the speedup compared to the single-threaded executions. Another noise factor is the nondeterministic nature of the system from the runtime point of view. We repeated every test ten times and calculated the average of our measurements to mitigate these interfering factors and to make the results more precise and robust.

The following tables show the number of function evaluations and the time they took in milliseconds as a function of the hardness and the number of threads. An increase in the number of function evaluations was found. The overhead that we experienced was proportional to the number of threads, due to the frequent need of synchronizations during the optimization, and inversely proportional to the computation time of the objective function. You can see that the runtimes decrease in case of every function up to application of four threads even without the alteration of the evaluation time by the hardness. This remains true for up to eight threads in most of the cases. The optimization time may increase by the addition of further threads to the execution. We call this phenomenon the parallel saturation of the optimization. This happens when the synchronization overhead of the threads overcomes the gain of parallel execution. Additional threads only make things worse in such cases. The data series without hardness for the *Easom* function demonstrate well this phenomenon.

This saturation starts to manifest if we use a greater number of threads, and we set the function evaluation time at most ten times longer. Fitting a curve on the runtimes, we can observe that its minimum is translated toward a higher number of applied threads. This means that more threads are required for saturation as the overhead ratio is lowered with the longer function evaluations. The runtime is continuously decreasing up to using 16 threads if we apply a 100 or 1000 multiplier to the function evaluation times. We could not reach the saturation point with these tests. The *Easom* and *Shubert* functions have a lesser runtime by one magnitude than the other ones. The optimizer already found the optimum after approximately 10^4 function evaluations for these two problems, while roughly 10^5 evaluations were needed for the other cases. The optimizer probably did not find the known global optimum in the latter cases, but that is irrelevant from the multithreading speedup point of view. The difference of magnitudes of function evaluations points out that the parallelization of the computation is not affected when the evaluation numbers are high enough (Table 4.1).

Table 4.1 Ackley, Easom, and Levy test function results

Time factor	Threads	Ackley		Easom		Levy	
		NFE	Runtime (ms)	NFE	Runtime (ms)	NFE	Runtime (ms)
1x	1	100,447	3553.7	10,120.7	122.7	101,742	3245.8
	2	101,544	2216.0	10,246.4	118.4	104,827	2062.1
	4	102,881	1515.3	10,506.2	112.0	112,351	1473.0
	8	102,908	1145.9	11,078.7	145.0	129,056	1218.9
	16	110,010	1319.3	12,335.9	149.0	156,907	1548.4
10x	1	100,553	5838.5	10,141.9	165.0	102,412	5414.6
	2	101,510	3370.6	10,273.7	132.0	106,339	3057.4
	4	103,495	2014.7	10,524.9	111.6	114,325	2100.8
	8	105,977	1480.0	11,096.6	135.7	126,848	1592.3
	16	112,008	1623.1	12,308.2	157.6	155,884	1714.4
100x	1	100,516	27,868.7	10,117.7	413.2	102,227	25,364.9
	2	101,585	14,544.5	10,256.9	352.5	106,423	13,788.8
	4	103,420	7806.3	10,546.1	323.9	115,158	8030.5
	8	107,657	4544.7	11,083.6	296.3	130,109	5368.2
	16	115,264	3648.3	12,313.7	257.8	167,983	5205.0
1000x	1	100,567	258,690.0	10,198.5	1722.4	102,028	236,066.5
	2	101,561	126,315.0	10,249.7	865.1	106,792	123,220.0
	4	103,616	68,691.5	10,521.7	508.7	114,847	70,399.6
	8	107,718	39,430.4	11,116.0	441.6	128,450	45,762.8
	16	115,875	27,021.5	12,358.5	389.3	160,364	37,713.4

The number of function evaluations is increased due to the parallel computation induced by information loss. This means only 0.5–1.5% per thread, given the configuration we used. The explanation is that the local searches do not stop after the global evaluation limit is reached; thus every local search that is started right before this limit will potentially use all the locally allowed number of function evaluations. This behavior can be prevented if the local searches consider the global limit of allowed function evaluations, but this has not been implemented in the current optimization framework (Table 4.2).

Table 4.2 Rastrigin-20, Schwefel-6, and Shubert test function results

Time factor	Threads	Rastrigin-20		Schwefel-6		Shubert	
		NFE	Runtime (ms)	NFE	Runtime (ms)	NFE	Runtime (ms)
1x	1	100,479	2869.3	100,155	3027.4	10,085.1	145.0
	2	100,921	1881.3	100,252	2337.6	10,192.2	140.2
	4	101,436	1413.2	100,470	1555.4	10,381.4	128.0
	8	103,476	1252.7	101,205	1596.5	10,788.4	127.2
	16	107,937	1256.9	102,285	1408.3	11,686.9	145.2
10x	1	100,328	4201.9	100,189	3484.9	10,087.0	207.6
	2	101,015	2514.4	100,362	2334.2	10,181.5	164.9
	4	102,049	1682.6	100,616	1738.2	10,375.3	143.3
	8	104,993	1364.4	100,691	1586.7	10,842.9	141.8
	16	106,210	1444.7	101,445	1487.2	11,778.9	149.8
100x	1	100,354	17,438.0	100,185	7524.7	10,091.1	836.5
	2	101,396	9114.6	100,327	4656.3	10,171.7	534.3
	4	102,800	4958.8	100,782	2858.5	10,370.2	347.8
	8	106,106	3053.4	101,176	2298.8	10,805.3	289.2
	16	112,977	2627.7	102,331	1932.1	11,718.0	246.4
1000x	1	100,485	135,399.0	10,0248	44,046.3	10,091.4	6436.0
	2	101,293	70,335.2	10,0394	22,780.3	10,177.3	3375.9
	4	103,041	37,275.3	10,1183	12,169.5	10,383.5	1847.3
	8	106,244	21,411.8	10,2259	7329.0	10,775.7	1249.9
	16	113,135	14,740.0	10,4775	5466.7	11,679.3	979.4

We ran the optimization procedures with the following configuration:

```
<?xml version="1.0"?>
<Global package="org.uszeged.inf.optimization.algorithm"
    class="optimizer.global.serialized.SerializedGlobal">
  <ThreadCount type="long">
    NUMCORES
  </ThreadCount>
  <NewSampleSize type="long">
    10000
  </NewSampleSize>
  <SampleReducingFactor type="double">
    0.66666
  </SampleReducingFactor>
  <LocalOptimizer class="optimizer.local.parallel.NUnirandiCLS">
    <MaxFunctionEvaluations type="long">
      10000
    </MaxFunctionEvaluations>
```

```
    <RelativeConvergence type="double">
        0.00000001
    </RelativeConvergence>
    <LineSearchFunction
        class="optimizer.line.parallel.LineSearchImpl">
    </LineSearchFunction>
  </LocalOptimizer>
  <Clusterizer class="clustering.serialized.
SerializedGlobalSingleLinkageClusterizer">
    <Alpha type="double">
        0.01
    </Alpha>
  </Clusterizer>
</Global>
```

4.8.3 *SerializedGlobalSingleLinkageClusterizer Parallelization Test*

We tested the clustering module to study the effect of parallel computation to the running time of the clustering cycle. Although the clustering is completely independent from the underlying objective function, we wanted to study it on real data; therefore we chose to sample the five-dimensional *Rastrigin* function in the $x_i \in [-5, 5]$ interval. We generated N sample points to evaluate the function at these. We selected random pairs from these sample points and added them to the clustering module through the *clustering samples* algorithm (Algorithm 4.4).

This preparation resulted in approximately $N/2$ clusters of two elements. As a second step of the setup, we repeated the sample generation, but the new samples were loaded into the clusterizer algorithm together as unclustered elements. We started the multithreaded clustering in which all threads executed the parallel clustering procedure (Algorithm 4.2). When a thread exited the module, it was also stopped and could not reenter the clusterizer.

We measured how much time passed until all the threads stopped working on the second sample set. The presented data is the average of ten runs. You can observe in Table 4.3 that an improvement is achieved even in the case of low sample sizes. On the other hand, the clustering easily became saturated that happens much later when the sample size is greater. The unclustered samples are around 93% meaning that the number of executed comparisons is greater than 93% of the theoretical maximum, that is, at least $0.93 * N^2$.

Table 4.3 Clusterizer stress test results

Samples	Threads	Runtime (ms)	Unclustered samples
10^2	1	7.3	92.5
	2	6.1	95.5
	4	6.6	94.3
	8	8.8	95.4
	16	12.9	94.1
10^3	1	132.5	938.7
	2	94.5	938.7
	4	77.0	936.3
	8	80.6	940.7
	16	120.1	938.3
10^4	1	11,351.5	9372.5
	2	6255.4	9373.6
	4	3549.3	9369.7
	8	2730.1	9352.1
	16	2183.1	9354.8
10^5	1	1,996,087.9	93,685.0
	2	1,079,716.8	93,711.1
	4	533,583.0	93,690.8
	8	337,324.0	93,741.4
	16	224,280.0	93,698.7

4.8.4 Comparison of Global and PGlobal Implementations

The aim of the comparison is to reveal the differences between the Global and PGlobal implementations regarding the number of function evaluations. We applied the same configuration for both optimizers; we ran PGlobal on a single thread the same way as we ran Global. We studied 3 different local search algorithms and 63 test functions to have sufficient data. We ran every (global optimizer, local optimizer, test function) configuration 100 times to determine the necessary number of function evaluations to find the global optimum. We dropped any results of the 100 runs that were unsuccessful by using more than 10^5 function evaluations, and we calculated the average of the remaining values. We call the ratio of successful runs among all runs the robustness of the optimization. We only studied the configurations that had a 100% robustness for both global optimizers.

The following is the configuration of the optimizer algorithm parameters for this test:

```
<?xml version="1.0"?>
<Global package="org.uszeged.inf.optimization.algorithm"
    class="<global_optimizer>">
  <NewSampleSize type="long">
    400
```

```
  </NewSampleSize>
  <SampleReducingFactor type="double">
    0.03999
  </SampleReducingFactor>
  <LocalOptimizer
      class="optimizer.local.parallel.<local_optimizer>">
    <MaxFunctionEvaluations type="long">
      100000
    </MaxFunctionEvaluations>
    <RelativeConvergence type="double">
      0.00000001
    </RelativeConvergence>
    <LineSearchFunction
        class="optimizer.line.parallel.LineSearchImpl">
    </LineSearchFunction>
  </LocalOptimizer>
  <Clusterizer class="<clusterizer>">
    <Alpha type="double">
      0.01
    </Alpha>
  </Clusterizer>
</Global>
```

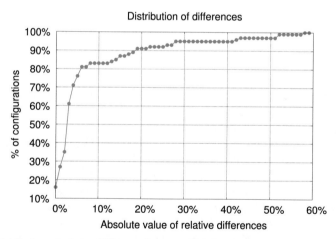

Fig. 4.8 Distribution of relative differences between Global and PGlobal

Figure 4.8 shows that in 80% of compared configurations, the relative difference was lower than 7%. The correlation between the two data vectors is 99.87%. The differences in the results are caused by many factors. The results were produced based on random numbers which can cause an error of a few percent. For the two optimizer processes, the data were generated in a different manner which can also cause uncertainties. These differences are hugely amplified by local optimization. If every local search would converge into the global optimum, the number of function evaluations would be dominated by one local search. In case of multiple local searches, the exact number is highly uncertain. With random starting points and not optimized step lengths, the local search converges to a nearly random local optima. The proportion of function evaluations will approximate the number of local searches; hence the number of function evaluations is unstable in these cases. We observed that on functions which have many local optima added to a slower function as noise, the differences are in the common range in contrast to the high differences that can be observed on "flat" and noisy functions. We suspect that the low noise behavior is caused by the implicit averaging of the gradients along the local search.

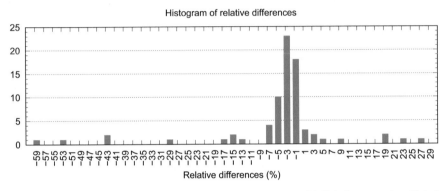

Fig. 4.9 Relative difference in the number of function evaluations of PGlobal compared to Global

On Figure 4.9 most of the relative error results are in the $[-7, 3]$ range. It shows a slight tendency that PGlobal uses on average less function evaluations. Most of the extreme values also favor PGlobal.

Finally, we have tested three local search methods whether the parallel use on a single core has different efficiency characteristic. The computational test results are summarized in Tables 4.4, 4.5, and 4.6, for the local search algorithms of NUnirandi, Unirandi, and Rosenbrock, respectively. The last column of these tables gives the relative difference between the number of function evaluations needed for the compared to implementations.

For all three local search techniques, we can draw the main conclusion that the old method and the serialized one do not differ much. In the majority of the cases, the relative difference is negligible, below a few percent. The average relative differences are also close to zero, i.e., the underlying algorithm variants are basically well balanced. The relative differences of the summed number of function evaluations needed by the three local search methods were -1.08%, -3.20%, and 0.00%, respectively.

Table 4.4 NUnirandiCLS results: what is the difference between the Serialized Global and Global in terms of number of function evaluations? The difference is negative when the parallel version is the better

Function	Serialized NFEV	Global NFEV	Difference
Beale	672.9	792.6	−15.12%
Booth	605.3	628.0	−3.61%
Branin	528.0	559.7	−5.65%
Cigar-5	1547.1	1518.1	1.91%
Colville	2103.4	2080.7	1.09%
Discus-40	27,131.8	27,586.0	−1.65%
Discus-rot-40	26,087.3	26,927.7	−3.12%
Discus-5	4943.0	5297.2	−6.69%
Discus-rot-5	4758.9	4940.1	−3.67%
Easom	1795.6	1708.8	5.08%
Ellipsoid-5	4476.1	4567.3	−2.00%
Griewank-20	8661.7	8005.2	8.20%
Hartman-3	605.7	644.6	−6.03%
Matyas	625.4	647.2	−3.37%
Rosenbrock-5	4331.5	3664.0	18.22%
Shubert	545.9	963.9	−43.37%
Six hump	502.9	524.9	−4.20%
Sphere-40	4169.8	4178.8	−0.22%
Sphere-5	827.2	853.1	−3.03%
Sum Squares-40	12,370.8	12,495.6	−1.00%
Sum Squares-5	881.0	915.9	−3.81%
Sum Squares-60	21,952.9	21,996.8	−0.20%
Trid	3200.8	3095.2	3.41%
Zakharov-40	17,958.9	18,334.0	−2.05%
Zakharov-5	984.7	1009.5	−2.46%
Average	6090.7	6157.4	−2.74%

Table 4.5 UnirandiCLS results: what is the difference between the Serialized Global and Global in terms of number of function evaluations? The difference is negative when the parallel version is the better

Function	Serialized NFEV	Global NFEV	Difference
Beale	762.3	1060.1	−28.10%
Booth	576.6	600.8	−4.02%
Branin	516.1	540.7	−4.55%
Discus-rot-40	33,824.2	35,055.3	−3.51%
Discus-5	18,605.9	19,343.3	−3.81%
Discus-rot-5	14,561.8	15,513.8	−6.14%
Goldstein Price	502.4	584.2	−13.99%
Griewank-20	9847.8	10,185.8	−3.32%
Matyas	615.2	646.1	−4.78%
Shubert	517.0	895.1	−42.24%
Six hump	480.6	501.0	−4.06%
Sphere-40	4083.1	4118.7	−0.86%
Sphere-5	781.9	794.3	−1.56%
Sum Squares-40	24,478.5	24,272.1	0.85%
Sum Squares-5	856.2	867.9	−1.35%
Zakharov-40	20,431.5	20,811.9	−1.83%
Zakharov-5	953.4	983.9	−3.10%
Average	7787.9	8045.6	−7.43%

Table 4.6 RosenbrockCLS results: what is the difference between the Serialized Global and Global in terms of number of function evaluations? The difference is negative when the parallel version is the better

Function	Serialized NFEV	Global NFEV	Difference
Beale	709.6	831.9	−14.70%
Booth	593.8	616.5	−3.68%
Branin	599.9	620.4	−3.30%
Six hump	546.5	563.5	−3.02%
Cigar-5	1536.9	1600.0	−3.94%
Cigar-rot-5	3438.6	3551.2	−3.17%
Colville	2221.4	2307.7	−3.74%
Discus-40	30,721.0	31,059.7	−1.09%
Discus-rot-40	30,685.2	30,960.3	−0.89%
Discus-5	2946.2	3113.2	−5.36%
Discus-rot-5	2924.3	3085.3	−5.22%
Discus-rot-60	47,740.6	48,086.6	−0.72%
Easom	4,664.2	11,178.8	−58.28%
Ellipsoid-5	4493.3	4509.8	−0.37%
Goldstein Price	569.2	693.0	−17.87%
Griewank-20	12,647.5	12,222.8	3.47%
Hartman-3	975.2	1040.7	−6.29%
Hartman-6	3047.7	2493.5	22.23%
Matyas	628.5	651.8	−3.58%
Powell-24	42,488.6	43,425.8	−2.16%
Powell-4	1950.9	2006.7	−2.78%
Rosenbrock-5	4204.7	3527.7	19.19%
Shekel-5	4790.0	3775.0	26.89%
Shubert	553.5	1153.3	−52.01%
Sphere-40	7788.6	7839.7	−0.65%
Sphere-5	905.0	924.2	−2.08%
Sum Squares-40	30,688.5	30,867.8	−0.58%
Sum Squares-5	970.1	1005.3	−3.50%
Sum Squares-60	71,891.9	72,063.5	−0.24%
Trid	3919.0	3925.0	y0.15%
Zakharov-40	34,177.5	35,605.0	−4.01%
Zakharov-5	1123.7	1178.1	−4.62%
Zakharov-60	80,742.9	82,393.3	−2.00%
Average	13,269.2	13,269.0	−4.19%

4.9 Conclusions

This chapter provided the considerations along which we have designed and implemented the parallel version of the GlobalJ algorithm. Our main aim was to have a code that is capable to utilize the widely available computer architectures that support efficient parallelization. The careful testing confirmed our expectations and proved that the parallel implementation of PGlobal can utilize multiple core computer architectures. For easy-to-solve problems with low computational cost, the PGlobal may show weaker efficiency. But for computationally expensive objective functions and for difficult to solve problems, the parallel version of Global can achieve closely linear speedup ratio, i.e., the total solution time can more or less be divided by the number of available CPU cores. The other way around, we have checked what are the costs of parallelization. According to our computational tests, the parallel implementation of the local search algorithms needed mostly somewhat less function evaluations than their serial use—when run on a single core.

Chapter 5
Example

5.1 Environment

Before we can start work with the GLOBAL optimizer package, we must set up a proper environment. The package uses the Java 8 virtual machine. To use the package with compiled objective functions, the Java 8 Runtime Environment (JRE) is sufficient. However, the common case is that the objective function is not compiled, and it implies the need for the Java 8 Development Kit (JDK). Both systems can be downloaded from *https://java.com/en/*.

5.2 Objective Function

We have to provide an objective function in the form of a Java class. The class must implement the *org.uszeged.inf.optimization.data.Function* interface thus all of its functions. The implementation must not have inner state if the optimization is done in parallelized environment.

The *boolean isDimensionAcceptable(int dim)* function receives the dimension setting before every optimization. If the setting is applicable for the objective function, it must return *true*; otherwise it returns *false*. For example, the $\sum_{i=1}^{dim} x_i^2$ can be easily extended for any positive integer dimensions. On the other hand, the function can be the simulation of a race car, and the highest velocity is tuned through the wheels radius. In this case there is no place for more dimensions.

The *boolean isParameterAcceptable(Vector lb, Vector ub)* function receives the lower bound of each dimension in *lb* and the upper bound of each dimension in *ub*. The optimization is bounded by this N dimensional rectangle. The function have to

© The Author(s), under exclusive licence to Springer International Publishing AG,
part of Springer Nature 2018
B. Bánhelyi et al., *The GLOBAL Optimization Algorithm*,
SpringerBriefs in Optimization, https://doi.org/10.1007/978-3-030-02375-1_5

return the value *true* if the bounds are acceptable and otherwise *false*. In the example of the race car, the radius must be positive and not bigger than some reasonable amount.

The last function is the *double evaluate(Vector x)* which calculates the function value at the given point *x*. The vector *x* is guaranteed to have every coordinate between the corresponding *lb* and *ub* values.

To keep this example simple, we choose an interesting but algebraically solvable function. It can be easily shown that the

$$(x_1 - 10)^2 * (\ln(x_1)^2 + 1) + x_2^2 * (\sin(x_2) + 1.1)$$

function has several local optima and a single global optima at $f(10,0) = 0$. The variable x_1 must be bigger than 0, and x_2 can be any rational number.

A Java implementation of the discussed functionality:

```java
// CustomFunction.java
import org.uszeged.inf.optimization.data.Function;
import org.uszeged.inf.optimization.data.Vector;

public class CustomFunction implements Function{

    private double sqr(double x){
        return x*x;
    }

    public boolean isDimensionAcceptable(int dim){
        return dim == 2;
    }

    public boolean isParameterAcceptable(Vector lb, Vector ub){
        return lb.getCoordinate(1) > 0;
    }

    public double evaluate(Vector x){
        double x1 = x.getCoordinate(1);
        double x2 = x.getCoordinate(2);

        double a = sqr(x1-10)*(sqr(Math.log(x1))+1);
        double b = sqr(x2)*(Math.sin(x2)+1.1d);

        return a+b;
    }

}
```

After finishing the function, save it to a directory and name it *CustomFunction.java*. Copy the *global.jar* compiled package to the same directory and open the command line. In the directory type for Windows *javac -cp .;global.jar CustomFunction.java* and *javac -cp .:global.jar CustomFunction.java* for linux. Now the objective function is compiled, and it can be used with the package.

5.3 Optimizer Setup

The optimizer can be set up both from the Java code and an XML file. It is convenient to create the optimizer structure and set the parameters from code, but it requires some experience with the package. Now we present the easier solution (for beginners).

First of all we have to choose the modules to optimize with. We will use the GlobalJ algorithm with the Unirandi local search method. GlobalJ is implemented by the *org.uszeged.inf.optimization.algorithm.optimizer.global.Global* class, and Unirandi is implemented by the

org.uszeged.inf.optimization.algorithm.optimizer.local.Unirandi

class. We use the built in clusterizer which is implemented by the

org.uszeged.inf.optimization.algorithm.clustering.GlobalSingleLinkageClusterizer

class.

Global has the following parameterization. The *NewSampleSize* parameter is set to 100; the *SampleReducingFactor* is 0.1. This means that Global generates 100 samples on every iteration, and then it selects the $100 * 0.1 = 10$ best samples for clustering. The *MaxNumberOfFunctionEvaluations* is used in a soft condition for all the function evaluations. It is only set to override the default value which is too low. The local optimizer type is a parameter of Global but it also has its own parameters. *MaxFunctionEvaluations* sets a limit for the maximum number of function evaluations during each of the local searches; the value is chosen to be 10,000. The *RelativeConvergence* parameter will determine the minimal slope that is considered to be nonzero; now its value is 10^{-8}. The clusterizer is also a parameter for Global, and its *Alpha* parameter determines the critical distance. Higher values of *Alpha* cause faster shrinking of the critical distance. The 0.2 value is about middle range.

The following XML file will be named *GlobalUnirandi.xml* and saved to the example directory.

The root nodes name must be *Global*. It has the optional *package* attribute which sets a basis package name. We must provide a *class* attribute which is the full name of the global optimizer class using the *package* as a prefix. The child nodes are parameters for the selected optimizer. Their name will be converted to a *setXYZ* function call. Primitive types have the *type* attribute to select the primitive type,

and the value will be converted respectively. Other classes have the *class* attribute
similar to the root and can have the *package*. The *package* will be overwritten only
in the nodes subtree where the root is the node.

```xml
<?xml version="1.0"?>
<Global package="org.uszeged.inf.optimization.algorithm"
    class="optimizer.global.Global">
 <NewSampleSize type="long">
   100
 </NewSampleSize>
 <SampleReducingFactor type="double">
   0.1
 </SampleReducingFactor>
 <MaxNumberOfFunctionEvaluations type="long">
   1000000
 </MaxNumberOfFunctionEvaluations>
 <LocalOptimizer class="optimizer.local.Unirandi">
   <MaxFunctionEvaluations type="long">
     10000
   </MaxFunctionEvaluations>
   <RelativeConvergence type="double">
     0.00000001
   </RelativeConvergence>
 </LocalOptimizer>
 <Clusterizer class="clustering.GlobalSingleLinkageClusterizer">
   <Alpha type="double">
     0.2
   </Alpha>
 </Clusterizer>
</Global>
```

5.4 Run the Optimizer

Before the optimization can be started, we have to define the boundaries. To
have some exciting parts in the optimization, let us choose the bounds to be
$lb = (0.1, -50)$, $ub = (20, 50)$. In this range there are multiple local optima and
the global optimum which is close to another local optima. The data goes into the
CustomFunction.bnd file as follows:

```
CustomFunction
CustomFunction
2
0.1 20
-50 50
```

The file format consists of the printable function name, the Java class path from where it can be loaded, the dimension count, and the bounds for every dimensions. If the bounds are the same for every dimension, then the lower and upper bounds follow the dimension count, in separate lines each.

To run the optimizer with the previous settings, type the command

java -cp .;global.jar Calculate -f CustomFunction.bnd -o GlobalUnirandi.xml

If you use linux change the classpath to *.:global.jar*. The result should be four numbers, number of function evaluations, the run time in milliseconds, the optimum value, and some optimizer specific values. The present implementation of Global returns the number of local searches. Due to the random sampling and random local search techniques, the results will vary on every execution. In this case the typical values are between 300–1200 evaluations, 50–150 ms, 0–0.2 for the value, and 1–5 local searches. The *Alpha* value of 0.9 results in much higher robustness and evaluation count. It varies then from 3000 to 15,000 evaluations.

5.5 Constraints

The optimizer package has the ability to handle the bound constraints on the search space. However, there are a large number of use cases when nonlinear constraints are required. The package itself does not have any direct support for it, but there is a common technique to overcome the lack this functionality. We can introduce a penalty function that has a constant value higher than the original objective function's possible maximum value in the area. The distance from the target area is added to the constant value, and this sum will represent the function value for the outside region. If the evaluation point is inside the constrained area, the penalty term is not taken into account. The use of the penalty function approach in connection to GLOBAL algorithm has been discussed in detail in [13] together with some theoretical statements on the reliability of the obtained solution and computational result on its use to prove the chaotic behavior of certain nonlinear mappings.

Let us see an example with our custom objective function. Let the constrained area be the circle centered at $O(5,4)$ with radius 6. This circle violates the interpretation region of the function, so we take its intersection with the bounding box $x_1 \in [0.1, 11]$, $x_2 \in [-2, 10]$. The constant value has to be bigger than $620 + 130 = 750$; to be sure we choose it an order of magnitude higher, let it be 10,000. We intentionally choose the constraints such that the global optimum is outside of the valid region.

Since the optimization methods did not change, the optimizer configuration XML can be reused from the previous example. The *ConstrainedFunction1.java* implementation ensures that the objective function is not evaluated outside the constrained region. The *isConstraintViolated(Vector x)* function returns the true value if the evaluation point is outside of the constrained region.

```
// ConstrainedFunction1.java
import org.uszeged.inf.optimization.data.Function;
import org.uszeged.inf.optimization.data.Vector;
import org.uszeged.inf.optimization.util.VectorOperations;

public class ConstrainedFunction1 implements Function{

    private static final Vector center = new Vector(new
        double[]{5, 4});
    private static final double radius = 6;
    private static final double penaltyConstant = 10000d;

    public boolean isConstraintViolated(Vector x){
        Vector diff = VectorOperations.subtractVectors(x, center);
        return Vector.norm(diff) > radius;
    }

    private double sqr(double x){
        return x*x;
    }

    public boolean isDimensionAcceptable(int dim){
        return dim == 2;
    }

    public boolean isParameterAcceptable(Vector lb, Vector ub){
        return lb.getCoordinate(1) > 0;
    }

    public double evaluate(Vector x){

        if (isConstraintViolated(x)){
            Vector diff = VectorOperations.subtractVectors(x, center);
            return Vector.norm(diff) + penaltyConstant;
        } else {
            double x1 = x.getCoordinate(1);
            double x2 = x.getCoordinate(2);

            double a = sqr(x1-10)*(sqr(Math.log(x1))+1);
            double b = sqr(x2)*(Math.sin(x2)+1.1d);

            double originalValue = a+b;

            return originalValue;
        }
    }
}
```

The corresponding BND file is *Constrained1.bnd*.

```
ConstrainedFunction1
ConstrainedFunction1
2
0.1 11
-2 10
```

Compile the file with the *javac -cp .;global.jar ConstrainedFunction1.java* command, and run it with the *java -cp .;global.jar Calculate -f Constrained1.bnd -o GlobalUnirandi.xml* command. The optimum value should be about 0.4757, and the run time should be much less than a second.

There is a special case for this kind of constraining that can be used sometimes. If the objective function is interpreted in a such search box that it contains the constrained area, it is advised to use a different approach. The penalty function remains the same as described earlier. It has 0 value inside the constrained region. The optimizer will search the optimum on the sum of the objective function and the penalty function.

To try the special approach of constraints, compile the following file. You can notice that now the original objective function is evaluated every time, and the value is used alone or with the penalty.

```
// ConstrainedFunction2.java
import org.uszeged.inf.optimization.data.Function;
import org.uszeged.inf.optimization.data.Vector;
import org.uszeged.inf.optimization.util.VectorOperations;

public class ConstrainedFunction2 implements Function{

    private static final Vector center = new Vector(new
        double[]{5, 4});
    private static final double radius = 6;
    private static final double penaltyConstant = 10000d;

    public boolean isConstraintViolated(Vector x){
        Vector diff = VectorOperations.subtractVectors(x, center);
        return Vector.norm(diff) > radius;
    }

    private double sqr(double x){
        return x*x;
    }

    public boolean isDimensionAcceptable(int dim){
        return dim == 2;
    }

    public boolean isParameterAcceptable(Vector lb, Vector ub){
        return lb.getCoordinate(1) > 0;
    }
```

```
public double evaluate(Vector x){
    double x1 = x.getCoordinate(1);
    double x2 = x.getCoordinate(2);

    double a = sqr(x1-10)*(sqr(Math.log(x1))+1);
    double b = sqr(x2)*(Math.sin(x2)+1.1d);

    double originalValue = a+b;

    if (isConstraintViolated(x)){
        Vector diff = VectorOperations.subtractVectors(x, center);
        return originalValue + Vector.norm(diff) +
            penaltyConstant;
    } else {
        return originalValue;
    }
}
}
```

The corresponding BND file is the following.

```
ConstrainedFunction2
ConstrainedFunction2
2
0.1 11
-2 10
```

Compile the function again, run it with the same configuration like before, and check the results. The optimum value should be the same, and the function evaluation count can sometimes drop below the first versions best values, because of the better quality information.

If the constrained area is one dimensional, it is recommended to create a new base variable. In our example we choose a new t variable. The constraints for the variables are

$$x_1 = t^3$$

$$x_2 = t^5$$

We can substitute x_1 and x_2 into the original function, and it can be optimized like any other function. After the optimization the results can easily be transformed into the original problem space:

$$(t^3 - 10)^2 * (\ln(t^3)^2 + 1) + (t^5)^2 * (\sin(t^5) + 1.1)$$

5.6 Custom Module Implementation

To implement a custom module, first we have to specify the algorithm and the parameters. Our example is a local optimizer module which can be run on multiple instances. The algorithm is a simple hill climbing method on a raster grid which can change size during the search. It starts with a grid and checks its neighbors in every direction along the axes. If there is a neighbor that has lower value than the base point, that neighbor becomes the base point. If all neighbors have larger values, the algorithm performs a magnitude step down; it divides the current step length with 2.33332. After that it continues the search on the grid.

There are three conditions that can cause the optimizer to stop. The relative convergence measures the difference between the last two function values. If the relative decrease is lower than the convergence value, the algorithm stops. The step length is also bounded by this parameter. If the step length is lower than the convergence value, the algorithm stops. When a step size decrease occurs, the algorithm checks if the maximum number of such step downs is reached and exits if necessary. The last condition is on the number of function evaluations.

Algorithm 5.1 Discrete climber

Input

> *starting point*
> *configuration*

Return value

> *optimum*

```
 1: while true do
 2:      generate neighbors and check for new optimum
 3:      if new optimum found then
 4:          if relative convergence limit exceeded then
 5:              break
 6:          end if
 7:          set new point as base point
 8:      else
 9:          if magnitude step down limit exceeded then
10:              break
11:          else
12:              perform magnitude step down
13:          end if
14:      end if
15:      if maximum number of function evaluations exceeded then
16:          break
17:      end if
18: end while
19: store optimum
```

The implementation follows the guidelines present in the package. A key feature is the *Builder* pattern which helps to instantiate optimizer modules. The Builder is a nested class of the module; therefore, it can access and set up the inner state. The Builder object has many setter functions to receive the parameters. The *build( )* function checks the validity of the parameter set, saves the parameters in the log file, and returns the fully parameterized module instance. The objective function-related parameters, such as starting point, bounds, and the objective function itself, are loaded through setter functions into the module. The main class, included in the optimizer package, assembles the optimization modules and controls the optimization from the setup phase to displaying the results.

For simplicity we can reuse our custom objective function implemented by *CustomFunction.java*. The *CustomFunction.bnd* will also fit our needs. The only thing left is the optimizer configuration. We want to use the *Global* optimizer and our new *DiscreteClimber* local optimizer. The new configuration is identical with the *GlobalUnirandi.xml* except the local optimizer module. Notice that the *LocalOptimizer* tag has its *package* attribute set to the empty string. This is necessary because the implementation of the *DiscreteClimer* class is in the default package, not in the *org.uszeged.inf.optimization.algorithm* package.

```xml
<?xml version="1.0"?>
<Global package="org.uszeged.inf.optimization.algorithm"
    class="optimizer.global.Global">
 <NewSampleSize type="long">
  100
 </NewSampleSize>
 <SampleReducingFactor type="double">
  0.1
 </SampleReducingFactor>
 <MaxNumberOfFunctionEvaluations type="long">
  1000000
 </MaxNumberOfFunctionEvaluations>
 <LocalOptimizer package="" class="DiscreteClimber">
  <MaxFunctionEvaluations type="long">
   1000
  </MaxFunctionEvaluations>
  <MaxMagnitudeStepDowns type="long">
   32
  </MaxMagnitudeStepDowns>
  <RelativeConvergence type="double">
   0.00000001
  </RelativeConvergence>
 </LocalOptimizer>
 <Clusterizer class="clustering.GlobalSingleLinkageClusterizer">
  <Alpha type="double">
   0.9
  </Alpha>
 </Clusterizer>
</Global>
```

Save the configuration file, and compile the local optimizer module with the *javac -cp .;global.jar DiscreteClimber.java* command. Run the optimization with the *java -cp .;global.jar Calculate -f CustomFunction.bnd -o GlobalDiscrete.xml* command. The optimization should finish quickly with around 3500 function evaluations. With this configuration the global optimum is found robustly. The Discrete-Climber code can be found in Appendix C.

Appendix A
User's Guide

This chapter presents the parametrization of the PGlobal modules and how they depend on each other.

The Builder design pattern is responsible for the parametrization of the modules. The module classes have a statically enclosed Builder class which can access all the variables in the module instances. The *Builder* class defines setter functions for the parameters and stores the data in a *Configuration* object. The *Builder* also defines a *build()* function to instantiate the module, load with custom and default parameters, and then return the valid parameterized object.

For simplicity, we are going to use the shorthand ...*xyz.Abc* for the long class name of *org.uszeged.inf.optimization.algorithm.xyz.Abc* in the explanations. The type of the parameters are going to be between square brackets before the parameter names; the parameters marked with *(required)* must be provided.

A.1 Global Module

The module Global is the previously discussed GlobalJ implementation. The interface ...*optimizer.global.GlobalOptimizer* defines its functionality. It is implemented in the ...*optimizer.global.Global* class.

A.1.1 Parameters

- **[module] Clusterizer (required):** a clustering module must be provided that implements the interface ...*clustering.Clusterizer<Point>*.
- **[module] LocalOptimizer (required):** a local optimizer must be provided that implements the interface ...*optimizer.local.LocalOptimizer<Vector>*

B. Bánhelyi et al., *The GLOBAL Optimization Algorithm*, SpringerBriefs in Optimization, https://doi.org/10.1007/978-3-030-02375-1

- **[long] MaxNumberOfSamples:** Maximum number of sample points generated during the whole optimization process.
- **[long] NewSampleSize:** Number of sample points generated in one iteration.
- **[double] SampleReducingFactor:** Denotes the portion of *NewSampleSize* to be selected for clustering.
- **[long] MaxNumberOfIterations:** Maximum number of main optimization cycles.
- **[long] MaxNumberOfFunctionEvaluations:** Maximum number of function evaluations during the whole optimization process. It is a soft condition because the local optimization will not stop at the global optimizer limit.
- **[long] MaxNumberOfLocalSearches:** Maximum number of local searches, checked at the end of the main optimization cycle.
- **[long] MaxNumberOfLocalOptima:** Maximum number of local optima.
- **[long] MaxRuntimeInSeconds:** The maximum runtime of the optimization, checked at the end of the main optimization cycle.
- **[double] KnownGlobalOptimumValue:** A special parameter to help benchmark tests of the optimizer, checked after every local optimization.

A.2 SerializedGlobal Module

The module SerializedGlobal is the previously discussed implementation of the PGlobal algorithm. The interface ... *optimizer.global.GlobalOptimizer* defines its functionality. It is implemented in the
... *optimizer.global.serialized.SerializedGlobal* class.

A.2.1 Parameters

- **[module] Clusterizer (required):** a clustering module must be provided that implements the interface ... *clustering.serialized.SerializedClusterizer<Point>*.
- **[module] LocalOptimizer (required):** a local optimizer module must be provided that implements the interface
... *optimizer.local.parallel.ParallelLocalOptimizer<Vector>*.
- **[long] MaxNumberOfSamples:** Denotes the maximum number of sample points generated during the whole optimization process.
- **[long] NewSampleSize:** Number of sample points generated for one iteration.
- **[double] SampleReducingFactor:** Denotes the portion of *NewSampleSize* to be selected for clustering.
- **[long] MaxNumberOfIterations:** Number of iterations is defined to be the number of clustering cycles. The maximum number of iterations matches the maximum number of clustering cycles.

- **[long] MaxFunctionEvaluations:** Maximum number of function evaluations during the whole optimization process. It is a soft condition because the local optimization will not stop at the global optimizer limit and the thread handling can cause overshoot.
- **[long] MaxNumberOfLocalSearches:** Maximum number of local searches, overshoot can occur due to thread handling.
- **[long] MaxNumberOfLocalOptima:** Maximum number of local optima found before threads start to exit.
- **[long] MaxRuntimeInSeconds:** The maximum runtime of the optimization, overshoot can occur due to thread handling and local searches.
- **[double] KnownGlobalOptimumValue:** A special parameter to help benchmark tests of the optimizer.
- **[long] ThreadCount:** Number of optimizer threads.
- **[long] LocalSearchBatchSize:** Denotes the number of sample points transferred from the clusterizer to the local optimizer after each clustering attempt. If this value is lower than the number of sample points in the clusterizer, then some sample points will stay in place. If the batch size is set to be 0, an adaptive algorithm will set the batch size to the number of available threads.

A.3 GlobalSingleLinkageClusterizer Module

This module is responsible for the clustering of N-dimensional points. It is the only member of the package that implements the interface

...clustering.Clusterizer<Point>

ensuring the usability of the module Global. It is implemented in the class

...clustering.GlobalSingleLinkageClusterizer.

A.3.1 Parameters

- **[double] Alpha (required):** Determines the size function of the critical distance. N is the sum of clustered and unclustered sample points; n is the dimension of the input space:

$$d_c = \left(1 - \alpha^{\frac{1}{N-1}}\right)^{\frac{1}{n}}, \ \alpha \in [0, 1]$$

With lower *Alpha* the critical distance shrinks slower.

A.4 SerializedGlobalSingleLinkageClusterizer Module

The module, similarly to the GlobalSingleLinkageClusterizer, is responsible for the clustering of *N*-dimensional points. It implements the interface

... *clustering.serialized.SerializedClusterizer<Point>.*

Its internal operation supports the multi-threaded environment of SerializedGlobal. It is implemented in the class

... *clustering.serialized.SerializedGlobalSingleLinkageClusterizer.*

A.4.1 Parameters

– **[double] Alpha (required):** Determines the size function of the critical distance. *N* is the sum of clustered and unclustered sample points; *n* is the dimension of the input space:

$$d_c = \left(1 - \alpha^{\frac{1}{N-1}}\right)^{\frac{1}{n}}, \ \alpha \in [0, 1]$$

With lower *Alpha* the critical distance shrinks slower.

A.5 UNIRANDI Module

UNIRANDI is a local search algorithm based on random walk. The module provides a complete functionality; there is no need for any additional modules to be able to use it. It implements the interface ... *optimizer.local.LocalOptimizer<Vector>*. It is implemented in the ... *optimizer.local.Unirandi* class.

A.5.1 Parameters

– **[double] InitStepLength:** Initial step length of the algorithm. Smaller initial step lengths can increase the number of function evaluations and the probability of staying in the region of attraction.
– **[string] DirectionRandomization:** Selects the direction randomization method. The *UNIT_CUBE* setting generates a normalized vector using independent uniform distributions for each dimension. If it is set to *NORMAL_DISTRIBUTION*, then it will generate a normalized vector from the uniform distribution on the surface of a hypersphere.

- **[long] MaxFunctionEvaluations:** Maximum number of function evaluations during the local search. This is a soft condition; overshoot can occur due to the line search method.
- **[double] RelativeConvergence:** Determines the minimum step length and the minimum decrease in value between the last two points.

A.6 NUnirandi Module

NUnirandi is a local search algorithm also based on random walk. It is the improved version of the UNIRANDI algorithm. The module provides a complete functionality; there is no need for any additional modules to be able to use it. It implements the interface *... optimizer.local.LocalOptimizer<Vector>*. It is implemented in the *... optimizer.local.NUnirandi* class.

A.6.1 Parameters

- **[double] InitStepLength:** Initial step length of the algorithm. Smaller initial step lengths can increase the number of function evaluations and the probability of staying in the region of attraction.
- **[string] DirectionRandomization:** Selects the direction randomization method. The *UNIT_CUBE* setting generates a normalized vector using independent uniform distributions for each dimension. If it is set to *NORMAL_DISTRIBUTION*, then it will generate a normalized vector from the uniform distribution on the surface of a hypersphere.
- **[long] MaxFunctionEvaluations:** Maximum number of function evaluations during the local search. This is a soft condition; overshoot can occur due to the line search method.
- **[double] RelativeConvergence:** Determines the minimum step length and the minimum decrease in value between the last two points.

A.7 UnirandiCLS Module

The module is a variant of Unirandi. UnirandiCLS must be provided a line search algorithm in contrast to the original algorithm that has a built-in one. The rest of the parametrization is the same. It implements the interface

... optimizer.local.parallel.ParallelLocalOptimizer<Vector>.

It is implemented in the *... optimizer.local.parallel.UnirandiCLS* class.

A.7.1 Parameters

- **[module] LineSearchFunction (required):** Line search module that implements the *...optimizer.line.parallel.ParallelLineSearch<Vector>* interface.
- **[double] InitStepLength:** Initial step length of the algorithm. Smaller initial step lengths can increase the number of function evaluations and the probability of staying in the region of attraction.
- **[string] DirectionRandomization:** Selects the direction randomization method. The *UNIT_CUBE* setting generates a normalized vector using independent uniform distributions for each dimension. If it is set to *NORMAL_DISTRIBUTION*, then it will generate a normalized vector from the uniform distribution on the surface of a hypersphere.
- **[long] MaxFunctionEvaluations:** Maximum number of function evaluations during the local search. This is a soft condition; overshoot can occur due to the line search method.
- **[double] RelativeConvergence:** Determines the minimum step length and the minimum decrease in value between the last two points.

A.8 NUnirandiCLS Module

The module is a variant of NUnirandi. NUnirandiCLS must be provided a line search algorithm in contrast to the original algorithm that has a built-in one. The rest of the parametrization is the same. It implements the interface

... optimizer.local.parallel.ParallelLocalOptimizer<Vector>.

It is implemented in the *... optimizer.local.parallel.NUnirandiCLS* class.

A.8.1 Parameters

- **[module] LineSearchFunction (required):** Line search module that implements the *...optimizer.line.parallel.ParallelLineSearch<Vector>* interface.
- **[double] InitStepLength:** Initial step length of the algorithm. Smaller initial step lengths can increase the number of function evaluations and the probability of staying in the region of attraction.
- **[string] DirectionRandomization:** Selects the direction randomization method. The *UNIT_CUBE* setting generates a normalized vector using independent uniform distributions for each dimension. If it is set to *NORMAL_DISTRIBUTION*, then it will generate a normalized vector from the uniform distribution on the surface of a hypersphere.

- **[long] MaxFunctionEvaluations:** Maximum number of function evaluations during the local search. This is a soft condition; overshoot can occur due to the line search method.
- **[double] RelativeConvergence:** Determines the minimum step length and the minimum decrease in value between the last two points.

A.9 Rosenbrock Module

The module implements the Rosenbrock local search method. It implements the *...optimizer.local.parallel.ParallelLocalOptimizer<Vector>* interface. It is implemented in the *... optimizer.local.parallel.Rosenbrock* class.

A.9.1 Parameters

- **[module] LineSearchFunction (required):** Line-search module that implements the *...optimizer.line.parallel.ParallelLineSearch<Vector>* interface.
- **[double] InitStepLength:** Initial step length of the algorithm. Smaller initial step lengths can increase the number of function evaluations and the probability of staying in the region of attraction.
- **[long] MaxFunctionEvaluations:** Maximum number of function evaluations during the local search. This is a soft condition; overshoot can occur due to the line search method.
- **[double] RelativeConvergence:** Determines the minimum step length and the minimum decrease in value between the last two points.

A.10 LineSearchImpl Module

The module implements the *...optimizer.line.parallel.ParallelLineSearch< Vector >* interface. The module is the Unirandi's built-in line search algorithm. Hence, the running only depends on the starting point and the actual step length of the local search; there are no parameters. The algorithm is walking with doubling steps until the function value starts to increase. It is implemented in the class

... optimizer.line.parallel.LineSearchImpl.

Appendix B
Test Functions

In this appendix we give the details of the global optimization test problems applied for the computational tests. For each test problem, we give the full name, the abbreviated name, the dimension of the problem, the expression of the objective function, the search domain, and the place and value of the global minimum.

- Name: Ackley function
  Short name: Ackley
  Dimensions: 5
  Function:

$$f(x_1,x_2,x_3,x_4,x_5) = -20\exp\left(-0.2\sqrt{\frac{1}{5}\sum_{i=1}^{5}x_i^2}\right) - \exp\left(\frac{1}{5}\sum_{i=1}^{5}\cos(2\pi x_i)\right)$$
$$+20+\exp(1)$$

  Search domain:
$$-15 \le x_1,\ldots,x_d \le 30$$

  Global minimum:
$$f(3,0.5) = 0$$

- Name: Beale's function
  Short name: Beale
  Dimensions: 2
  Function:

$$f(x_1,x_2) = (1.5-x_1+x_1x_2)^2 + \left(2.25-x_1+x_1x_2^2\right)^2 + \left(2.625-x_1\left(1-x_1^3\right)\right)^2;$$

  Search domain:
$$-4.5 \le x_1,x_2 \le 4.5$$

© The Author(s), under exclusive licence to Springer International Publishing AG, part of Springer Nature 2018
B. Bánhelyi et al., *The GLOBAL Optimization Algorithm*,
SpringerBriefs in Optimization, https://doi.org/10.1007/978-3-030-02375-1

Global minimum:

$$f(3,0.5) = 0$$

- Name: Booth Function
  Short name: Booth
  Dimensions: 2
  Function:

$$f(x_1,x_2) = (x_1 + 2x_2 - 7)^2 + (2x_1 + x_2 - 5)^2$$

Search domain:

$$-10 \leq x_1, x_2 \leq 10$$

Global minimum:

$$f(0,0) = 0$$

- Name: Branin function
  Short name: Branin
  Dimensions: 2
  Function:

$$f(x_1,x_2) = \left(x_2 - \frac{5.1}{4\pi^2}x_1^2 + \frac{5}{\pi}x_1 - 6 \right)^2 + 10 \left(1 - \frac{1}{8\pi} \right) \cos(x_1) + 10$$

Search domain:

$$-5 \leq x_1, x_2 \leq 15$$

Global minimum: $f(-\pi, 12.275) = 0.3978873577,$ $f(\pi, 2.275)$
$= 0.3978873577$, and $f(9.42478, 2.675) = 0.3978873577$.

- Name: Cigar function
  Short name: Cigar-5, Cigar-40, Cigar-rot-5, Cigar-rot-40, Cigar-rot-60[1]
  Dimensions: 5, 40, 60
  Function[2]:

$$f(x_1,\ldots,x_d) = x_1^2 + 10^3 \sum_{i=2}^{d} x_i^2$$

Search domain:

$$-5 \leq x_1,\ldots,x_d \leq 5$$

Global minimum:

$$f(0,\ldots,0) = 0$$

- Name: Colville function
  Short name: Colville
  Dimensions: 4

---

[1] Rotation versions.
[2] In MATLAB: 10^4 instead of 10^3.

Function:

$$f(x_1, x_2, x_3, x_4) = 100(x_1^2 - x_2)^2 + (x_1 - 1)^2 + (x_3 - 1)^2 + 90(x_3^2 - x_4)^2$$
$$+ 10.1((x_2 - 1)^2 + (x_4 - 1)^2) + 19.8(x_2 - 1)(x_4 - 1)$$

Search domain:

$$-10 \le x_1, x_2, x_3, x_4 \le 10$$

Global minimum:

$$f(1, 1, 1, 1) = 0$$

- Name: Sum of different powers function
  Short name: Diff. powers-5, diff. powers-40, diff. powers-60
  Dimensions: 5, 40, 60
  Function:

$$f(x_1, \ldots, x_d) = \sum_{i=1}^{d} |x|_i^{2+4\frac{i-1}{d-1}}$$

Search domain:

$$-5 \le x_1, \ldots, x_d \le 5$$

Global minimum:

$$f(0, \ldots, 0) = 0$$

- Name: Discus function
  Short name: Discus-5, Discus-40, Discus-rot-5, Discus-rot-40, Discus-rot-60
  Dimensions: 5, 40, 60
  Function:

$$f(x_1, \ldots, x_d) = 10^4 x_1^2 + \sum_{i=2}^{d} x_i^2$$

Search domain:

$$-5 \le x_1, \ldots, x_d \le 5$$

Global minimum:

$$f(0, \ldots, 0) = 0$$

- Name: Dixon-Price function
  Short name: Dixon-Price
  Dimensions: 10
  Function:

$$f(x_1, \ldots, x_10) = (x_1 - 1)^2 + \sum_{i=2}^{10} i \left(2x_i^2 - x_{i-1}\right)^2$$

Search domain:

$$-10 \le x_1, \ldots, x_d \le 10$$

Global minimum:

$$f(2^{-\frac{2^1-2}{2^1}}, 2^{-\frac{2^2-2}{2^2}}, \ldots, 2^{-\frac{2^{10}-2}{2^{10}}}) = 0$$

- Name: Easom function
  Short name: Easom
  Dimensions: 2
  Function:

$$f(x_1, x_2) = -\cos(x_1)\cos(x_2)\exp(-(x_1 - \pi)^2 - (x_2 - \pi)^2)$$

Search domain:
$$-100 \le x_1, x_2 \le 100$$

Global minimum:
$$f(\pi, \pi) = -1$$

- Name: Elipsoid function
  Short name: Elipsoid-5, Elipsoid-40, Elipsoid-rot-5, Elipsoid-rot-40, Elipsoid-rot-60
  Dimensions: 5, 40, 60
  Function:

$$f(x_1, \ldots, x_d) = \sum_{i=1}^{d} 10^{4\frac{i-1}{d-1}} x_i^2$$

Search domain:
$$-5 \le x_1, x_2 \le 5$$

Global minimum:
$$f(0, \ldots, 0) = 0$$

- Name: Goldstein Price function
  Short name: Goldstein-Price
  Dimensions:
  Function:[2]

$$f(x_1, x_2) = \left(1 + (x_1 + x_2 + 1)^2 (19 - 14x_1 + 3x_1^2 - 14x_2 + 6x_1 x_2 + 3x_2^2)\right)$$
$$\left(30 + (2x_1 - 3x_2)^2 (18 - 32x_1 + 12x_1^2 + 48x_2 - 36x_1 x_2 + 27x_2^2)\right)$$

Search domain:
$$-2 \le x_1, x_2 \le 2$$

Global minimum:
$$f(0, -1) = 3$$

- Name: Griewank function
  Short name: Griewank-5, Griewank-20
  Dimensions: 5, 20
  Function:

$$f(x_1, \ldots, x_d) = \sum_{i=1}^{d} \frac{x_i^2}{4000} - \prod_{i=1}^{d} \cos\left(\frac{x_i}{\sqrt{i}}\right) + 1$$

Search domain:
$$-10 \le x_1, x_2 \le 10$$

Global minimum:

$$f(0,\ldots,0) = 0$$

- Name: Hartman three-dimensional function
  Short name: Hartman-3
  Dimensions: 3
  Function:

$$f(x_1,x_2,x_3) = \sum_{i=1}^{4} \alpha_i \exp\left(-\sum_{j=1}^{3} A_{ij}(x_j - P_{ij})^2 \right),$$

where

$$\alpha = (1.0, 1.2, 3.0, 3.2)^T$$

$$A = \begin{bmatrix} 3.0 & 10 & 30 \\ 0.1 & 10 & 35 \\ 3.0 & 10 & 30 \\ 0.1 & 10 & 35 \end{bmatrix}$$

$$P = 10^{-5} \begin{bmatrix} 36890 & 11700 & 26730 \\ 46990 & 43870 & 74700 \\ 10910 & 87320 & 55470 \\ 3815 & 57430 & 88280 \end{bmatrix}$$

Search domain:

$$0.0 \le x_1, x_2, x_3 \le 1.0$$

Global minimum:

$$f(0.114614, 0.555649, 0.852547) = -3.8627821478$$

- Name: Hartman six-dimensional function
  Short name: Hartman-6
  Dimensions: 6
  Function:

$$f(x_1,x_2,x_3) = \sum_{i=1}^{4} \alpha_i \exp\left(-\sum_{j=1}^{6} A_{ij}(x_j - P_{ij})^2 \right),$$

where

$$\alpha = (1.0, 1.2, 3.0, 3.2)^T$$

$$A = \begin{bmatrix} 10 & 3 & 17 & 3.5 & 1.7 & 8 \\ 0.05 & 10 & 17 & 0.1 & 8 & 14 \\ 3 & 3.5 & 1.7 & 10 & 17 & 8 \\ 17 & 8 & 0.05 & 10 & 0.1 & 14 \end{bmatrix}$$

$$P = 10^{-4} \begin{bmatrix} 1312 & 1696 & 5569 & 124 & 8283 & 5886 \\ 2329 & 4135 & 8307 & 3736 & 1004 & 9991 \\ 2348 & 1451 & 3522 & 2883 & 3047 & 6650 \\ 4047 & 8828 & 8732 & 5743 & 1091 & 381 \end{bmatrix}$$

Search domain:

$$0.0 \leq x_1, x_2, x_3, x_4, x_5, x_6 \leq 1.0$$

Global minimum:

$$f(0.20169, 0.150011, 0.476874, 0.476874, 0.275332, 0.311652, 0.6573) =$$
$$-3.322368011415511$$

- Name: Levy function
  Short name: Levy
  Dimensions: 5
  Function:

$$f(x_1, \ldots, x_d) = \sin^2(\pi \omega_1) + \sum_{i=1}^{d-1} (\omega_i - 1) \left[1 + 10 \sin^2(\pi \omega_i + 1) \right] +$$
$$(\omega_d - 1)^2 \left[1 + \sin^2(2\pi \omega_d) \right],$$

  where

$$\omega_i = 1 + \frac{x_i - 1}{4}$$

  Search domain:

$$-10 \leq x_1, \ldots, x_d \leq 10$$

  Global minimum:

$$f(1, \ldots, 1) = 0$$

- Name: Matyas function
  Short name: Matyas
  Dimensions: 2
  Function:

$$f(x_1, x_2) = 0.26(x_1^2 + x_1^2) - 0.48 x_1 x_2$$

  Search domain:

$$-10 \leq x_1, x_2 \leq 10$$

  Global minimum:

$$f(0, 0) = 0$$

- Name: Perm-(d, β) function
  Short name: Perm-(4,1/2), Perm-(4,10)
  Dimensions: 4
  Function:

$$f(x_1, \ldots, x_d) = \sum_{i=1}^{d} \left(\sum_{j=1}^{d} (j^i + \beta) \left(\left(\frac{x_j}{j} \right)^i - 1 \right) \right)^2$$

Search domain:

$$-4 \leq x_1, \ldots, x_d \leq 4$$

Global minimum:

$$f(1, 2, \ldots, d) = 0$$

- Name: Powell function
  Short name: Powell-4, Powell-24
  Dimensions: 4, 24
  Function: $f(x_1, \ldots, x_d) = \sum_{i=1}^{d/4}[(x_{4i-3} + 10x_{4i-2})^2 + 5(x_{4i-1} - x_{4i})^2 + (x_{4i-2} - 2x_{4i-1})^4 + 10(x_{4i-3} - x_{4i})^4]$
  Search domain:

$$-4 \leq x_1, \ldots, x_d \leq 5$$

Global minimum:

$$f(0, 0, 0, 0) = 0$$

- Name: Power sum function
  Short name: Power sum
  Dimensions: 4
  Function:

$$f(x_1, \ldots, x_d) = \sum_{k=1}^{d} \left[\left(\sum_{i=1}^{d} x_i^k \right) - b_k \right]^2,$$

where

$$b = (8, 18, 44, 114)$$

Search domain:

$$0 \leq x_1, \ldots, x_d \leq 4$$

Global minimum:

$$f(1, 2, \ldots, d) = 0$$

- Name: Rastrigin function
  Short name: Rastrigin
  Dimensions: 4
  Function:

$$f(x_1, \ldots, x_d) = 10d + \sum_{i=1}^{d} \left[x_1^2 - 10\cos(2\pi x_i) \right]$$

Search domain:

$$-5.12 \leq x_1, \ldots, x_d \leq 5.12$$

Global minimum:

$$f(0, \ldots, 0) = 0$$

- Name: Rosenbrock function
  Short name: Rosenbrock-5, Rosenbrock-40, Rosenbrock-rot-5, Rosenbrock-rot-40, Rosenbrock-rot-60
  Dimensions: 5, 40, 60

Function:

$$f(x_1, \ldots, x_d) = \sum_{i=1}^{d} \left(100 \left(x_{i+1} - x_i^2 \right)^2 + (x_i - 1)^2 \right)$$

Search domain:

$$-10 \leq x_1, \ldots, x_d \leq 10$$

Global minimum:

$$f(1, \ldots, 1) = 0$$

- Name:　Schaffer function
  Short name:　Schaffer
  Dimensions:　2
  Function:

$$f(x_1, x_2) = 0.5 + \frac{\sin^2 \left(x_1^2 - x_2^2 \right) - 0.5}{\left[1 + 0.001 \left(x_1^2 + x_2^2 \right) \right]^2}$$

Search domain:

$$-20 \leq x_1, x_2 \leq 20$$

Global minimum:

$$f(0,0) = 0$$

- Name:　Schwefel function
  Short name:　Schwefel
  Dimensions:　5
  Function:

$$f(x_1, \ldots, x_d) = 418.9829d - \sum_{i=1}^{d} x_i \sin \left(\sqrt{|x_i|} \right)$$

Search domain:

$$-500 \leq x_1, \ldots, x_d \leq 500$$

Global minimum:

$$f(420.9687, \ldots, 420.9687) = 6.363918737406493 \ 10^{-05}$$

- Name:　Shekel function
  Short name:　Shekel-5, Shekel-7, Shekel-10
  Dimensions:　4
  Function:

$$f(x_1, x_2, x_3, x_4) = -\sum_{i=1}^{m} \left(\sum_{j=1}^{4} (x_j - C_{ji})^2 + c_i \right)^{-1},$$

where

$$m = 5,7,10$$

$$c = [0.1, 0.2, 0.2, 0.4, 0.4, 0.6, 0.3, 0.7, 0.5, 0.5]$$

$$C = \begin{bmatrix} 4 & 1 & 8 & 6 & 3 & 2 & 5 & 8 & 6 & 7 \\ 4 & 1 & 8 & 6 & 7 & 9 & 3 & 1 & 2 & 3.6 \\ 4 & 1 & 8 & 6 & 3 & 2 & 5 & 8 & 6 & 7 \\ 4 & 1 & 8 & 6 & 7 & 9 & 3 & 1 & 2 & 3.6 \end{bmatrix}$$

Search domain:

$$0 \leq x_1, x_2, x_3, x_4 \leq 10$$

Global minimum: $f_{m=5}(4,4,4,4) = -10.153199679058231$, $f_{m=7}(4,4,4,4) = -10.402940566818664$, and $f_{m=10}(4,4,4,4) = -10.536409816692046$

- Name: Sharp ridge function
  Short name: Sharpridge-5, Sharpridge-40
  Dimensions: 5, 40
  Function:

$$f(x_1, \ldots, x_d) = x_1^2 + 100\sqrt{\sum_{i=2}^{d} x_i^2}$$

Search domain:

$$-5 \leq x_1, x_2 \leq 5$$

Global minimum:

$$f(0, \ldots, 0) = 0$$

- Name: Shubert function
  Short name: Shubert
  Dimensions: 2
  Function:

$$f(x_1, x_2) = \left(\sum_{i=1}^{5} i \cos\left((i+1)x_1 + i\right) \right) \left(\sum_{i=1}^{5} i \cos\left((i+1)x_2 + i\right) \right)$$

Search domain:

$$-10 \leq x_1, x_2 \leq 10$$

Global minimum:

$$f(-5.12, 5.12) = -186.7309088310239$$

- Name: Six-hump camel function
  Short name: Six hump
  Dimensions: 2
  Function:

$$f(x_1,x_2) = \left(4 - 2.1x_1^2 + \frac{x_1^4}{3}\right)x_1^2 + x_1x_2 + \left(-4 + 4x_2^2\right)x_2^2$$

Search domain:

$$-3 \leq x_1,x_2 \leq 1$$

Global minimum: $f(0.0898, -0.7126) = -1.031628453$ and
$f(-0.0898, 0.7126) = -1.031628453$

- Name: Sphere function
  Short name: Sphere-5, Sphere-40
  Dimensions: 5, 40
  Function:

$$f(x_1,\ldots,x_d) = \sum_{i=1}^{d} x_i^2$$

Search domain:

$$-5 \leq x_1,\ldots,x_d \leq 5$$

Global minimum:

$$f(0,\ldots,0) = 0$$

- Name: Sum of squares function
  Short name: Sum squares-5, sum squares-40, sum squares-60, sum squares-rot-60
  Dimensions: 5, 40, 60
  Function:

$$f(x_1,\ldots,x_d) = \sum_{i=1}^{d} ix_i^2$$

Search domain:

$$-5 \leq x_1,x_2,x_3 \leq 5$$

Global minimum:

$$f(0,\ldots,0) = 0$$

- Name: Trid function
  Short name: Trid
  Dimensions: 10
  Function:

$$f(x_1,\ldots,x_d) = \sum_{i=1}^{d} (x_i - 1)^2 - \sum_{i=2}^{d} (x_i x_{i-1})$$

Search domain:

$$-100 \leq x_1,\ldots,x_d \leq 100$$

Global minimum:

$$f(0,\ldots,0) = -210$$

- Name: Zakharov function
  Short name: Zakharov-5, Zakharov-40, Zakharov-60, Zakharov-rot-60

Dimensions: 5, 40, 60
Function:

$$f(x_1,\ldots,x_d) = \sum_{i=1}^{d} x_i^2 + \left(\sum_{i=1}^{d} 0.5 i x_i \right)^2 + \left(\sum_{i=1}^{d} 0.5 i x_i \right)^4$$

Search domain:

$$-5 \le x_1,\ldots,x_d \le 10$$

For the rotated version:

$$-5 \le x_1,\ldots,x_d \le 5$$

Global minimum:

$$f(0,\ldots,0) = 0$$

Appendix C
DiscreteClimber Code

In this appendix we list the code of the local search procedure DisreteClimber used in the Chapter 5.

```
// DiscreteClimber
import org.uszeged.inf.optimization.algorithm.optimizer.
 OptimizerConfiguration;
import org.uszeged.inf.optimization.data.Vector;
import org.uszeged.inf.optimization.algorithm.optimizer.local.
 parallel.AbstractParallelLocalOptimizer;
import org.uszeged.inf.optimization.util.Logger;
import org.uszeged.inf.optimization.util.ErrorMessages;

public class DiscreteClimber extends
    AbstractParallelLocalOptimizer<Vector>{

  public static final String PARAM_MAX_MAGNITUDE_STEPDOWNS =
      "MAX_MAGNITUDE_STEPDOWNS";
  private static final long DEFAULT_MAX_FUNCTION_EVALUATIONS
      = 1000L;
  private static final long DEFAULT_MIN_FUNCTION_EVALUATIONS
      = 100L;
  private static final long DEFAULT_MAX_MAGNITUDE_STEPDOWNS =
      5L;
  private static final double DEFAULT_RELATIVE_CONVERGENCE =
      1E-12d;
  private static final double DEFAULT_MIN_INIT_STEP_LENGTH =
      0.001d;
  private static final double DEFAULT_MAX_INIT_STEP_LENGTH =
      0.1d;

  // It's better to have numbers that can be represented by
      fractions
  // with high denominator values and the number should be
      around 2.
  public static final double STEPDOWN_FACTOR = 2.33332d;
```

```
private double initStepLength;
private long maxMagnitudeStepDowns;
private long maxFunctionEvaluations;
private double relativeConvergence;

private double stepLength;
private Vector basePoint;
private double baseValue;
private Vector newPoint;
private double newValue;
private long dimension;
private long magnitudeStepDowns;
private boolean newPointFound;

private DiscreteClimber(){
  super();
}

public void reset(){
  super.reset();
}

public void restart(){
  super.restart();

  basePoint = new Vector(super.startingPoint);
  baseValue = super.startingValue;
  dimension = basePoint.getDimension();

  stepLength = initStepLength;
  magnitudeStepDowns = 0;
  numberOfFunctionEvaluations = 0;

  super.optimum = new Vector(basePoint);
  super.optimumValue = baseValue;
}

public void run(){
  if (!isRunnable) {
    Logger.error(this,"run() optimizer is not parameterized
        correctly");
    throw new IllegalArgumentException(
          ErrorMessages.LOCAL_NOT_PARAMETERIZED_YET);
  }

  while(true){

    // minimize neighbors
    newPointFound = false;
    newValue = baseValue;

    for (int i = 1; i <= dimension; i++){
      double value;
      double coordinateBackup = basePoint.getCoordinate(i);
```

```
        basePoint.setCoordinate(i, coordinateBackup +
            stepLength);
        value = objectiveFunction.evaluate(basePoint);
        numberOfFunctionEvaluations++;
        if (value < newValue){
          newValue = value;
          newPoint = new Vector(basePoint);
          newPointFound = true;
        }

        basePoint.setCoordinate(i, coordinateBackup -
            stepLength);
        value = objectiveFunction.evaluate(basePoint);
        numberOfFunctionEvaluations++;
        if (value < newValue){
          newValue = value;
          newPoint = new Vector(basePoint);
          newPointFound = true;
        }

        basePoint.setCoordinate(i, coordinateBackup);
      }

      if (newPointFound){
        // new point found in current magnitude

        // check if step length or decrease in function
            value is big enough
        if (Math.abs(stepLength) < relativeConvergence
            || (baseValue - newValue) / Math.abs(newValue) <
               relativeConvergence){
          Logger.trace(this,"run() exit condition: relative
              convergence");
          break;
        }

        basePoint = newPoint;
        baseValue = newValue;
      } else {
        // in current magnitude an optimum is reached

        // try step down , if the limit reached exit
        if (magnitudeStepDowns < maxMagnitudeStepDowns){
          magnitudeStepDowns++;
          stepLength /= STEPDOWN_FACTOR;
        } else {
          Logger.trace(this,"run() exit condition: magnitude
              step downs");
          break;
        }
      }

      // check if the function evaluation count is exceeded
```

```
      if (numberOfFunctionEvaluations >=
          maxFunctionEvaluations){
        Logger.trace(this,"run() exit condition: number of
            function evaluations");
        break;
      }
    }

    // save the optimum point to the conventional variables
    optimum.setCoordinates(basePoint.getCoordinates());
    optimumValue = baseValue;

    Logger.trace(this,"run() optimum: {0} : {1}",
      String.valueOf(super.optimumValue),
      super.optimum.toString()
      );
  }

  // Creates an exact copy of optimizer with link copy
  public DiscreteClimber getSerializableInstance(){
    Logger.trace(this,"getSerializableInstance() invoked");
    DiscreteClimber obj = (DiscreteClimber)
        super.getSerializableInstance();
    // Elementary variables are copied with the object itself
    // We need to copy the variables manually which extends
        Object class
    obj.basePoint = new Vector(basePoint);
    obj.newPoint = new Vector(newPoint);
    return obj;
  }

  public static class Builder {

    private DiscreteClimber discreteClimber;
    private OptimizerConfiguration<Vector> configuration;

    public Builder() {
      this.configuration = new
          OptimizerConfiguration<Vector>();
    }

    public void setInitStepLength(double stepLength) {

      if (stepLength < DEFAULT_MIN_INIT_STEP_LENGTH) {
        stepLength = DEFAULT_MIN_INIT_STEP_LENGTH;
      } else if (stepLength > DEFAULT_MAX_INIT_STEP_LENGTH) {
        stepLength = DEFAULT_MAX_INIT_STEP_LENGTH;
      }
      this.configuration.addDouble(PARAM_INIT_STEP_LENGTH,
          stepLength);
    }

    public void setMaxMagnitudeStepDowns(long stepDowns){
```

```
    if (stepDowns < 0) {
      stepDowns = 0;
    }
    this.configuration.addLong(PARAM_MAX_MAGNITUDE_STEPDOWNS,
        stepDowns);
  }

  public void setMaxFunctionEvaluations(long
      maxEvaluations) {

    if (maxEvaluations < DEFAULT_MIN_FUNCTION_EVALUATIONS) {
      maxEvaluations = DEFAULT_MIN_FUNCTION_EVALUATIONS;
    }
    this.configuration.addLong(PARAM_MAX_FUNCTION_EVALUATIONS,
        maxEvaluations);
  }

  public void setRelativeConvergence(double convergence) {

    if (convergence < DEFAULT_RELATIVE_CONVERGENCE) {
      convergence = DEFAULT_RELATIVE_CONVERGENCE;
    }
    this.configuration.addDouble(PARAM_RELATIVE_CONVERGENCE,
        convergence);
  }

  public DiscreteClimber build(){

    discreteClimber = new DiscreteClimber();
    discreteClimber.configuration.addAll(configuration);

    if (!discreteClimber.configuration.containsKey
        (PARAM_INIT_STEP_LENGTH)){
      discreteClimber.configuration.addDouble
        (PARAM_INIT_STEP_LENGTH,
            DEFAULT_MAX_INIT_STEP_LENGTH);
    }
    discreteClimber.initStepLength =
        discreteClimber.configuration.getDouble(
      PARAM_INIT_STEP_LENGTH);
    Logger.info(this,"build() INIT_STEP_LENGTH = {0}",
      String.valueOf(discreteClimber.initStepLength));

    if (!discreteClimber.configuration.containsKey
        (PARAM_MAX_MAGNITUDE_STEPDOWNS)) {
      discreteClimber.configuration.addLong
        (PARAM_MAX_MAGNITUDE_STEPDOWNS,
        DEFAULT_MAX_MAGNITUDE_STEPDOWNS);
    }
    discreteClimber.maxMagnitudeStepDowns =
        discreteClimber.configuration.
        getLong(PARAM_MAX_MAGNITUDE_STEPDOWNS);
    Logger.info(this,"build() MAX_MAGNITUDE_STEPDOWNS =
        {0}",
```

```
        String.valueOf(discreteClimber.maxMagnitude
            StepDowns));

    if (!discreteClimber.configuration.containsKey
        (PARAM_MAX_FUNCTION_EVALUATIONS)){
      discreteClimber.configuration.addLong
          (PARAM_MAX_FUNCTION_EVALUATIONS,
        DEFAULT_MAX_FUNCTION_EVALUATIONS);
    }
    discreteClimber.maxFunctionEvaluations =
        discreteClimber.configuration.getLong(
      PARAM_MAX_FUNCTION_EVALUATIONS);
    Logger.info(this,"build() MAX_FUNCTION_EVALUATIONS =
        {0}",
      String.valueOf(discreteClimber.maxFunction
          Evaluations));

    if (!discreteClimber.configuration.containsKey
        (PARAM_RELATIVE_CONVERGENCE)){
      discreteClimber.configuration.addDouble
          (PARAM_RELATIVE_CONVERGENCE,
        DEFAULT_RELATIVE_CONVERGENCE);
    }
    discreteClimber.relativeConvergence =
        discreteClimber.configuration.getDouble(
      PARAM_RELATIVE_CONVERGENCE);
    Logger.info(this,"build() RELATIVE_CONVERGENCE = {0}",
      String.valueOf(discreteClimber.relativeConvergence));

    return discreteClimber;
  }
 }
}
```

References

1. Apache Commons Math: http://commons.apache.org/proper/commons-math (2017)
2. Balogh, J., Csendes, T., Stateva, R.P.: Application of a stochastic method to the solution of the phase stability problem: cubic equations of state. Fluid Phase Equilib. **212**, 257–267 (2003)
3. Balogh, J., Csendes, T., Rapcsák, T.: Some Global Optimization Problems on Stiefel Manifolds. J. Glob. Optim. **30**, 91–101 (2004)
4. Banga, J.R., Moles, C.G., Alonso, A.A.: Global optimization of Bioprocesses using Stochastic and hybrid methods. In: C.A. Floudas, P.M. Pardalos (eds.) Frontiers in Global Optimization, pp. 45–70. Springer, Berlin (2003)
5. Bánhelyi, B., Csendes, T., Garay, B.M.: Optimization and the Miranda approach in detecting horseshoe-type chaos by computer. Int. J. Bifurcation Chaos **17**, 735–747 (2007)
6. Betró, B., Schoen, F.: Optimal and sub-optimal stopping rules for the multistart algorithm in global optimization. Math. Program. **57**, 445–458 (1992)
7. Boender, C.G.E., Rinnooy Kan, A.H.G.: Bayesian stopping rules for multistart global optimization methods. Math. Program. **37**, 59–80 (1987)
8. Boender, C.G.E., Rinnooy Kan, A.H.G.: On when to stop sampling for the maximum. J. Glob. Optim. **1**, 331–340 (1991)
9. Boender, C.G.E., Romeijn, H.E.: Stochastic methods. In: Horst, R., Pardalos, P. (eds.) Handbook of Global Optimization, pp. 829–869. Kluwer, Dordrecht (1995)
10. Boender, C.G.E., Zielinski, R.: A sequential Bayesian approach to estimating the dimension of a multinominal distribution. In: Sequential Methods in Statistics. Banach Center Publications, vol. 16. PWN-Polish Scientific Publisher, Warsaw (1982)
11. Boender, C.G.E., Rinnooy Kan, A.H.G., Timmer, G., Stougie, L.: A stochastic method for global optimization. Math. Program. **22**, 125–140 (1982)

B. Bánhelyi et al., *The GLOBAL Optimization Algorithm*,
SpringerBriefs in Optimization, https://doi.org/10.1007/978-3-030-02375-1

12. Csendes, T.: Nonlinear parameter estimation by global optimization-efficiency and reliability. Acta Cybernet. **8**, 361–370 (1988)
13. Csendes, T., Garay, B.M., Bánhelyi, B.: A verified optimization technique to locate chaotic regions of Hénon systems. J. Glob. Optim. **35**, 145–160 (2006)
14. Csendes, T., Bánhelyi, B., Hatvani, L.: Towards a computer-assisted proof for chaos in a forced damped pendulum equation. J. Comput. Appl. Math. **199**, 378–383 (2007)
15. Csendes, T., Pál, L., Sendin, J.O.H., Banga, J.R.: The GLOBAL optimization method revisited. Optim. Lett. **2**, 445–454 (2008)
16. Csete, M., Szekeres, G., Bánhelyi, B., Szenes, A., Csendes, T., Szabo, G.: Optimization of Plasmonic structure integrated single-photon detector designs to enhance absorptance. In: Advanced Photonics 2015, JM3A.30 (2015)
17. Currie, J., Wilson, D.I.: OPTI: Lowering the barrier between open source optimizers and the industrial MATLAB user. In: Sahinidis, N., Pinto, J. (eds.) Foundations of Computer-Aided Process Operations. Savannah, Georgia (2012)
18. Custódio, A.L., Rocha, H., Vicente, L.N.: Incorporating minimum Frobenius norm models in direct search. Comput. Optim. Appl. **46**, 265–278 (2010)
19. Davidon, W.: Variable metric method for minimization. Technical Report ANL5990 (revised), Argonne National Laboratory, Argonne, Il (1959)
20. Dolan, E., Moré, J.J.: Benchmarking optimization software with performance profiles. Math. Program. **91**, 201–213 (2002)
21. Grippo, L., Rinaldi, F.: A class of derivative-free nonmonotone optimization algorithms employing coordinate rotations and gradient approximations. Comput. Optim. Appl. **60**(1), 1–33 (2015)
22. Grishagin, V.A.: Operational characteristics of some global search algorithms. Prob. Stoch. Search **7**, 198–206 (1978)
23. Hansen, N., Auger, A., Finck, S., Ros, R.: Real-parameter black-box optimization benchmarking 2010: Experimental setup. Technical Report RR-7215, INRIA (2010)
24. Hansen, N., Auger, A., Ros, R., Finck, S., Posik, P.: Comparing results of 31 algorithms from the black-box optimization benchmarking BBOB-2009. In: GECCO'10: Proc. 12th Ann. Conf. on Genetic and Evolutionary Computation, pp. 1689–1696. ACM, New York (2010)
25. Hendrix, E.M.T., G.-Tóth, B.: Introduction to Nonlinear and Global Optimization. Optimization and its Application. Springer, Berlin (2010)
26. Hooke, R., Jeeves, T.A.: Direct search solution of numerical and statistical problems. J. ACM **8**, 212–226 (1961)
27. Horst, R., Pardalos, P.M. (eds.): Handbook of Global Optimization. Kluwer, Dordrecht (1995)
28. Huyer, W., Neumaier, A.: Global optimization by multilevel coordinate search. J. Glob. Optim. **14**, 331–355 (1999)
29. Järvi, T.: A random search optimizer with an application to a max-min problem. Publications of the Institute for Applied Mathematics (3). University of Turku, Finland (1973)

30. Johnson, S.: The NLopt nonlinear-optimization package. http://ab-initio.mit.edu/nlopt. Last accessed July 2015
31. JScience: http://jscience.org (2017)
32. JSGL: http://jgsl.sourceforge.net (2017)
33. JQuantLib: http://www.jquantlib.org (2017)
34. Kearfott, R.B.: Rigorous Global Search: Continuous Problems. Kluwer, Dordrecht (1996)
35. Kelley, C.T.: Detection and remediation of stagnation in the Nelder-Mead algorithm using a sufficient decrease condition. Siam J. Optim. **10**, 43–55 (1997)
36. Kelley, C.T.: Iterative Methods for Optimization. SIAM, Philadelphia (1999)
37. Locatelli, M., Schoen, F.: Random linkage: a family of acceptance/rejection algorithms for global optimization. Math. Program. **2**, 379–396 (1999)
38. Locatelli, M., Schoen, F.: Global Optimization: Theory, Algorithms, and Applications. MOS-SIAM Series on Optimization. SIAM, Philadelphia (2013)
39. Markót, M.Cs., Csendes, T.: A new verified optimization technique for the "packing circles in a unit square" problems. SIAM J. Optim. **16**, 193–219 (2005)
40. Mockus, J.: Bayesian Approach to Global Optimization. Kluwer, Dordrecht (1989)
41. Moles, C.G., Gutierrez, G., Alonso, A.A., Banga, J.R.: Integrated process design and control via global optimization – A wastewater treatment plant case study. Chem. Eng. Res. Des. **81**, 507–517 (2003)
42. Moles, C.G., Banga, J.R., Keller, K.: Solving nonconvex climate control problems: pitfalls and algorithm performances. Appl. Soft Comput. **5**, 35–44 (2004)
43. Montes de Oca, M.A., Aydin, D., Stützle, T.: An incremental particle swarm for large-scale continuous optimization problems: an example of tuning-in-the-loop (re)design of optimization algorithms. Soft Comput. **15**(11), 2233–2255 (2011)
44. Moré, J.J., Wild, S.M.: Benchmarking derivative-free optimization algorithms. SIAM J Optim. **20**, 172–191 (2009)
45. Murtagh, F., Contreras, P.: Algorithms for hierarchical clustering: an overview. Wiley Interdisc. Rew.: Data Min. Knowl. Disc. **2**, 86–97 (2012)
46. Nelder, J., Mead, R.: The downhill simplex method. Comput. J. **7**, 308–313 (1965)
47. NumPy: http://www.numpy.org (2017)
48. Pál, L.: Empirical study of the improved UNIRANDI local search method. Cen. Eur. J. Oper. Res. **25**(2017), 929–952 (2017). https://doi.org/10.1007/s10100-017-0470-2
49. Pál, L., Csendes, T.: An improved stochastic local search method in a multistart framework. In: Proceedings of the 10th Jubilee IEEE International Symposium on Applied Computational Intelligence and Informatics, Timisoara, pp. 117–120 (2015)
50. Pál, L., Csendes, T., Markót, M.Cs., Neumaier, A.: Black-box optimization benchmarking of the GLOBAL method. Evol. Comput. **20**, 609–639 (2012)

51. Pintér, J.D.: Global Optimization in Action. Kluwer, Dordrecht (1996)
52. Pošík, P., Huyer, W.: Restarted Local Search Algorithms for Continuous Black Box Optimization. Evol. Comput. **20**(4), 575–607 (2012)
53. Pošík, P., Huyer, W., Pál, L.: A comparison of global search algorithms for continuous black box optimization. Evol. Comput. **20**(4), 509–541 (2012)
54. Powell, M.J.D.: An efficient method for finding the minimum of a function of several variables without calculating derivatives. Comput. J. **7**(2), 155–162 (1964)
55. Powell, M.J.D.: The NEWUOA software for unconstrained optimization without derivatives. In: Di Pillo, G., Roma, M. (eds.) Large Scale Nonlinear Optimization, pp. 255–297. Springer, Berlin (2006)
56. Press, W.H., Teukolsky, S.A., Vetterling, W.T., Flannery, B.P.: Numerical Recipes in C. The Art of Scientific Computing, 2nd edn. Cambridge University Press, New York (1992)
57. PyQL: https://github.com/enthought/pyql (2017)
58. Rastrigin, L.A.: Random Search in Optimization Problems for Multiparameter Systems. Defense Technical Information Center (1967)
59. Rinnooy Kan, A.H.G., Timmer, G.T.: Stochastic global optimization methods Part I: Clustering methods. Math. Program. **39**, 27–56 (1987)
60. Rinnooy Kan, A.H.G., Timmer, G.T.: Stochastic global optimization methods part II: Multi level methods. Math. Program. **39**, 57–78 (1987)
61. Rios, L.M., Sahinidis, N.V.: Rios, L.M., Sahinidis, N.V.: Derivative-free optimization: a review of algorithms and comparison of software implementations. J. Glob. Optim. **56**, 1247–1293 (2013)
62. Rokach, L., Maimon, O.: Clustering Methods. Data Mining and Knowledge Discovery Handbook, pp. 321–352. Springer, New York (2005)
63. Rosenbrock, H.H.: An Automatic Method for Finding the Greatest or Least Value of a Function. Comput. J. **3**, 175–184 (1960)
64. SciPy: https://www.scipy.org (2017)
65. Sendín, J.O.H., Banga, J.R., Csendes, T.: Extensions of a Multistart Clustering Algorithm for Constrained Global Optimization Problems. Ind. Eng. Chem. Res. **48**, 3014–3023 (2009)
66. Sergeyev, Y.D., Kvasov, D.E.: Deterministic Global Optimization: An Introduction to the Diagonal Approach. Springer, New York (2017)
67. Sergeyev, Y.D., Strongin, R.G., Lera, D.: Introduction to Global Optimization Exploiting Space-Filling Curves. Springer Briefs in Optimization. Springer, New York (2013)
68. Szabó, P.G., Markót, M.Cs., Csendes, T., Specht, E., Casado, L.G., Garcia, I.: New Approaches to Circle Packing in a Square – With Program Codes. Springer, New York (2007)
69. The MathWorks, Inc.: https://www.mathworks.com/
70. TIOBE Index: https://www.tiobe.com/tiobe-index (2017)
71. Törn, A.A.: A search clustering approach to global optimization. In: Dixon, L., Szegő, G. (eds.) Towards Global Optimization, vol. 2, pp. 49–62. North-Holland, Amsterdam (1978)

72. Törn, A., Zilinskas, A.: Global Optimization. Lecture Notes in Computer Science, vol. 350. Springer, Berlin (1989)
73. WEKA: http://www.cs.waikato.ac.nz/ml/weka/index.html (2017)
74. Zhigljavsky, A.A., Zilinskas, A.: Stochastic Global Optimization. Springer, New York (2008)

Printed in the United States
By Bookmasters